Chemical Equilibria

**Chemical Thermodynamics Set**

coordinated by
Michel Soustelle

Volume 4

# Chemical Equilibria

Michel Soustelle

iSTE

WILEY

First published 2015 in Great Britain and the United States by ISTE Ltd and John Wiley & Sons, Inc.



ISTE Ltd
27-37 St George's Road
London SW19 4EU
UK

www.iste.co.uk

John Wiley & Sons, Inc.
111 River Street
Hoboken, NJ 07030
USA

www.wiley.com



Library of Congress Control Number: 2015951442

British Library Cataloguing-in-Publication Data
A CIP record for this book is available from the British Library
ISBN 978-1-84821-867-3

# Contents

# Preface

This book – an in-depth examination of chemical thermodynamics – is written for an audience of engineering undergraduates and Masters students in the disciplines of chemistry, physical chemistry, process engineering, materials, etc., and doctoral candidates in those disciplines. It will also be useful for researchers at fundamental- or applied-research labs dealing with issues in thermodynamics during the course of their work.

These audiences will, during their undergraduate degree, have received a grounding in general thermodynamics and chemical thermodynamics, which all science students are normally taught, and will therefore be familiar with the fundamentals, such as the principles and the basic functions of thermodynamics, and the handling of phase and chemical equilibrium states, essentially in an ideal medium, usually for fluid phases, in the absence of electrical fields and independently of any surface effects.

This set of books, which is positioned somewhere between an introduction to the subject and a research paper, offers a detailed examination of chemical thermodynamics that is necessary in the various disciplines relating to chemical or material sciences. It lays the groundwork necessary for students to go and read specialized publications in their different areas. It constitutes a series of reference books that touch on all of the concepts and methods. It discusses both scales of modeling: microscopic (by statistical thermodynamics) and macroscopic, and illustrates the link between them at every step. These models are then used in the study of solid, liquid and gaseous phases, either of pure substances or comprising several components.

The various volumes of the set will deal with the following topics:

– phase modeling tools: application to gases;

– modeling of liquid phases;

– modeling of solid phases;

– chemical equilibrium states;

– phase transformations;

– electrolytes and electrochemical thermodynamics;

– thermodynamics of surfaces, capillary systems and phases of small dimensions.

Appendices in each volume give an introduction to the general methods used in the text, and offer additional mathematical tools and some data.

This series owes a great deal to the feedback, comments and questions from all my students are the *Ecole nationale supérieure des mines* (engineering school) in Saint Etienne who have "endured" my lecturing in thermodynamics for many years. I am very grateful to them, and also thank them for their stimulating attitude. This work is also the fruit of numerous discussions with colleagues who teach thermodynamics in the largest establishments – particularly in the context of the group "Thermodic", founded by Marc Onillion. My thanks go to all of them for their contributions and conviviality.

This fourth instalment in the series is devoted to the study of chemical equilibria.

Chapter 1 describes transformations and chemical equilibria using Donder's affinity method. Equilibrium conditions are examined in enclosed media, where one or more equilibrium states are present, and in open systems. The chapter closes with a general look at azeotropic transformations.

Chapter 2 is a general study of the properties of physical and chemical equilibria. Thus, we examine the laws of displacement of equilibria under the influence of various disturbances, and the Gibbs and Duhem phase laws. Following a general study of indifferent states, the chapter goes on to analyze the conditions of stability of equilibria.

Chapter 3 discusses the different aspects of the law of mass action and the equilibrium constants associated therewith. A number of graphical representations used in the study of chemical equilibria are presented. In turn, we examine pole diagrams and equilibrium representations with temperature, with the generalization of Ellingham diagrams. The chapter ends with a presentation of binary, ternary and quaternary diagrams of chemical equilibria.

The fourth and final chapter in this volume is given over to the determination, both experimentally and by computation, of the values of the parameters associated with chemical reactions. Thermochemistry for enthalpy, the determination of the entropies, specific heat capacities and Gibbs energy values ultimately lead to the determination of the equilibrium constants. Analysis of the different thermodynamic tables and methods for estimating unknown values enable us to proceed to the practical application and finally computation of the equilibria by the equilibrium constant method and minimization of Gibbs energy.

Michel SOUSTELLE
Saint-Vallier,
September 2015

# Notations and Symbols

| | |
|---|---|
| $A$ : | area of a surface or an interface. |
| $\mathscr{A}$: | affinity. |
| A, B,: | components of a mixture. |
| $C$: | concentration. |
| $C_i$: | molar concentration (or molarity) of component $i$. |
| $C_V$, $C_P$: | specific heat capacity at constant volume and constant pressure, respectively. |
| $c$: | number of independent components. |
| $\mathrm{d}_eS$: | entropy exchange with the outside environment. |
| $\mathrm{d}_iS$: | internal entropy production. |
| $\mathrm{d}\omega$: | elementary volume. |
| $E$: | energy of the system. |
| $Eb$: | balance equation. |
| $\langle E \rangle$ : | mean total energy of an element in the canonical ensemble. |
| $E_C$: | total energy of the canonical ensemble. |
| $E_I$: | potential energy due to interactions. |
| $E_j$: | energy of an element $j$ of the canonical ensemble. |
| $E^0$: | standard emf of an electrochemical cell. |
| $\mathscr{E}_p$: | set of variables with $p$ intensive variables chosen to define a system. |
| $F$: | Helmholtz energy. |

| | |
|---|---|
| $F_m^{mix}$ : | molar excess Helmholtz energy. |
| $\overline{F}_i^{xs}$ : | partial molar excess Helmholtz energy of component *i*. |
| $\overline{F}_i^{mix}$ : | partial molar mixing Helmholtz energy of component *i*. |
| $\overline{F}_i$ : | free energy, partial molar Helmholtz energy of component *i*. |
| $F_m$: | molar Helmholtz energy. |
| $f_i$: | fugacity of the component *i* in a gaseous mixture. |
| $f_i^0$ : | molar Helmholtz energy of pure component *i*. |
| $f^0$ or $f_i^0$ : | fugacity of a pure gas *i*. |
| $\mathfrak{F}$: | Faraday (unit). |
| $G_m^{xs}$ : | excess Gibbs energy. |
| $\tilde{G}$ : | electrochemical Gibbs energy. |
| $\overline{G}_i^{xs}$ : | partial excess molar Gibbs energy of component *i*. |
| $G$, $\overline{G}_i$ , $[G]$: | Gibbs energy, partial molar Gibbs energy of *i*, generalized Gibbs energy. |
| $G_m$ : | molar Gibbs energy. |
| $G_m^{mix}$ : | molar Gibbs energy of mixing. |
| $g$: | degeneracy coefficient or multiplicity coefficient or statistical weight. |
| $g_i^0$ : | molar Gibbs energy of pure component *i*. |
| $g_i$: | coefficient of multiplicity of state *i*. |
| $g^*$: | molar Gibbs energy of gas *i* at pressure of 1 atmosphere in a mixture. |
| $H_T^0$ : | standard molar enthalpy of formation at temperature *T*. |
| $H$ , $\overline{H}_i$ : | enthalpy, partial molar enthalpy of *i*. |
| **H**: | Hamiltonian. |
| $\mathcal{H}$: | magnetic field. |
| $H_m^{xs}$ : | molar excess enthalpy. |
| $H_m^{mix}$ : | molar mixing enthalpy. |

| | |
|---|---|
| $\overline{H}_i^{xs}$: | partial excess molar enthalpy of component *i*. |
| $\overline{H}_i^{mix}$: | partial molar mixing enthalpy of component *i*. |
| h: | Planck's constant. |
| $h_i^0$: | molar enthalpy of pure component *i*. |
| $I, I_1, I_2, I_3$: | moments of inertia. |
| $I_t$: | integral of configuration of the canonical partition function of translation. |
| $\overline{J}_i$: | partial molar value of *J* relative to component *i*. |
| $J_i^{mix}$: | mixing value of *J* relative to component *i*. |
| $\overline{J}_i^{mix}$: | partial molar mixing value of *J* relative to component *i*. |
| $J_i^*$: | value of *J* relative to component *i* in a perfect solution. |
| $\overline{J}_i^*$: | partial molar value of *J* relative to component *i* in a perfect solution. |
| $j_i^0$: | value of *J* for the pure component *i* in the same state of segregation. |
| *j*: | rotational quantum number. |
| $K_{i,j}(\mathfrak{S}p)$: | thermodynamic coefficient associated with the set of variables $\mathfrak{S}p$. $X_j$ is its definition variable and $Y_i$ its definition function. |
| $K_i^{(\mathrm{Tr})}$: | constant of change of equilibrium for phase transition Tr for component *i*. |
| $K^{(c)}$: | equilibrium constant relative to concentrations. |
| $K^{(f)}$: | equilibrium constant relative to fugacity values. |
| $K^{(P)}$: | equilibrium constant relative to partial pressure values. |
| *K*: | equilibrium constant. |
| $k_B$: | Boltzmann's constant. |
| $L_t$: | latent heat accompanying transformation *t*. |
| *M*: | molar mass. |
| *m*: | total mass. |

$m_i$: mass of component *i*.

$N$: number of components of a solution or a mixture of gases or involved in a reaction or number of molecules of a collection.

$N_a$: Avogadro's number.

$N_A$: number of molecules of component A.

$N_C$: number of elements in the canonical ensemble.

$n_i$: number of objects *i* in the system with energy $\varepsilon_i$ or number of moles of component *i*.

$n$: translational quantum number or total number of moles in a solution or a mixture.

$n^{(\alpha)}$: total number of moles in a phase α.

$P$: pressure of a gas.

$P_i^{(subl)}$: sublimating vapor pressure of component *i*.

$P_i^{(vap)}$ $P_i^0$: saturating vapor pressure of component *i*.

$P_c$: critical pressure.

$P_i$: partial pressure of component *i.*

$p$: number of external physical variables.

$Q$: heat involved.

$Q_P$: heat of transformation at constant pressure.

$Q^{(P)}$: reaction quotient in terms of partial pressures.

$Q^{(I)}$: reaction quotient of a transformation in *reference state (I)*.

$Q_V$: transformation heat at constant volume.

$\mathfrak{R}$: reaction rate.

R: perfect gas constant.

$r_A$: radius of the ionic atmosphere.

$r_0$: minimum energy gap between two molecules.

$s_m^{mix}$: molar mixing entropy.

$s_m^{xs}$: partial excess molar entropy of component *i*.

$\dot{s}_m^{mix}$: partial mixing molar entropy of component *i*.

S: oversaturation of a solution.

$S_i$, $\bar{s}_i$: entropy or partial molar entropy of *i*.

| | |
|---|---|
| $\tilde{S}$ : | electrochemical entropy. |
| $s_m^{xs}$ : | excess molar entropy. |
| $s_i^0$ : | molar entropy of pure component $i$. |
| $T$: | temperature |
| $T_c^{mix}$ : | critical temperature of the mixture. |
| $T_c$: | critical temperature. |
| $T_F$: | Fermi temperature. |
| $T_i^{(Eb)}$: | boiling point of pure $i$. |
| $T_i^{(F)}$: | fusion (melting) point of pure $i$. |
| $T^{(sub)}$: | sublimation temperature. |
| $T^{(vap)}$: | vaporization temperature. |
| $T^{(\Delta\alpha)}$: | phase-change temperature. |
| $U_m^{xs}$ : | excess molar internal energy. |
| $U_m^{mix}$ : | mixing molar internal energy. |
| $\overline{U}_i^{xs}$ : | excess partial molar internal energy of component $i$. |
| $\overline{U}_i^{mix}$ : | partial mixing molar internal energy of component $i$. |
| $U$, $\overline{U}_i$ : | internal energy, partial molar internal energy of $i$. |
| $\tilde{U}$ : | internal electrochemical energy. |
| $U_m$: | molar internal energy. |
| $U_R$: | crosslink internal energy. |
| $u_i^0$ : | molar internal energy of pure component $i$. |
| $V$, $\overline{V}_i$ : | volume, partial molar volume of $i$. |
| $V_c$: | critical volume. |
| $V_G$: | Gibbs variance. |
| $V_m$: | molar volume. |
| $v_D$: | Duhem variance. |
| $v_i^0$ : | molar volume of pure component $i$. |
| v: | quantum vibration number. |
| $v_m$: | molecular volume. |

$v_M$: molar volume of solid at melting point.

$w_i$: mass fraction of component $i$.

$w_{ij}$: energy of exchange between atoms $i$ and $j$.

$x_k^{(\alpha)}$: molar fraction of component $k$ in phase $\alpha$.

$x, y, z$: coordinates of a point in space.

$x_i$: molar fraction of component $i$ in a solution.

$\langle y \rangle$: mean value of $y$.

$Y_i$ and $X_i$: conjugal intensive and extensive values.

$y_i$: molar fraction of component $i$ in a gaseous phase.

$Z_{AB}$: molecular partition function of interaction between molecules.

$Z_C$: canonical partition function.

$Z_{C(\mathrm{A})}$: canonical partition function of component A.

$Z_{C(I)}$: canonical partition function of interaction.

$Z_{C(t)}$: canonical partition function of translation.

$z$: molecular partition function.

$z_e$: electron molecular partition function or electrovalence of ion $i$.

$z_{int}$: contribution of internal motions to the molecular partition function.

$z_n$: molecular partition function of nuclei.

$z_{pf}$: molecular partition function of a perfect gas.

$z_r$: rotational molecular partition function.

$z_t$: translational molecular partition function.

$z_{t(pf)}$: translational molecular partition function of a perfect gas.

$z_v$: vibrational molecular partition function.

$\alpha$: Lagrange multiplier relating to the number of objects in a collection.

$\beta$: Lagrange multiplier relating to the energy of the objects in a collection.

$\Gamma(\mathcal{E}_P)$: characteristic function with the set $\mathcal{E}_P$ as canonical variables.

| | |
|---|---|
| $\Gamma_k$ : | coefficient of activity of a group. |
| $\Gamma$: | characteristic function. |
| $\gamma$: | coefficient of activity of the component *i* irrespective of the reference state. |
| $\gamma_0$: | activity coefficient of a solvent. |
| $\gamma_i$: | activity coefficient of the species *i* |
| $\gamma_i^{(\mathrm{I})}$ : | activity coefficient of component *i*, pure-substance reference. |
| $\gamma_i^{(\mathrm{II})}$ : | activity coefficient of component *i*, infinitely dilute solution reference. |
| $\gamma_i^{(\mathrm{III})}$ : | activity coefficient of component *i*, molar solution reference. |
| $\gamma_\pm$ : | mean activity coefficient of ions in an ionic solution. |
| $\gamma_\mathrm{S}$: | activity coefficient of a solute. |
| $\Delta_r A_T^0$ : | standard value at temperature *T* of *A* associated with the transformation *r*. |
| $\Delta_r(A)$: | value de *A* associated with the transformation *r*. |
| $\delta_{i,j}$: | Kronecker delta. |
| $\delta$: | coefficient of pressure increase at volume *V*. |
| $\varepsilon_{\mathrm{A(A)}}$: | lattice energy of an atom of A in lattice A. |
| $\varepsilon_{attr}$: | energy of attraction between molecules. |
| $\varepsilon_{\mathrm{C\text{-}H}}$: | energy of the C-H bond. |
| $\varepsilon_{i(e)}$: | electronic energy of a molecule *i*. |
| $\varepsilon_{i(I)}$: | interactional energy of a molecule *i*. |
| $\varepsilon_{i(n)}$: | nuclear energy of a molecule *i*. |
| $\varepsilon_{i(r)}$: | rotational energy of a molecule *i*. |
| $\varepsilon_{i(t)}$: | translational energy of a molecule *i*. |
| $\varepsilon_{i(v)}$: | vibrational energy of a molecule *i*. |
| $\varepsilon_{i,j}$: | energy of interaction between two molecules *i* and *j* or pair energy between atoms *i* and *j*. |
| $\varepsilon_p$: | potential energy of a molecule. |
| $\varepsilon_{rep}$: | repulsion energy between molecules. |

$\mu_i$: chemical potential of component $i$.

$\mu_i^{(\text{liq})}$, $\mu_i^{(\text{gas})}$ : chemical potential of component $i$ in liquid/gaseous state, respectively.

$\nu$: vibration frequency.

$\nu_{k(\rho)}$ : algebraic stoichiometric number of component $A_k$ in reaction $\rho$.

$\xi$: reaction extent.

$\rho$: density.

$\phi$: number of phases.

$\varphi_i$ : coefficient of fugacity of gas $i$ in a mixture.

$\phi^0$ or $\varphi_i^0$ : coefficient of fugacity of a pure gas.

$\Phi$: heat flux.

$\Theta_D$: Debye temperature.

$\Theta_E$: Einstein temperature.

$\omega_i$: set of position coordinates of molecule $i$.

# 1

# Physico-Chemical Transformations and Equilibria

A physico-chemical transformation of a system is the modification of the nature of its components under the influence of external stresses.

## 1.1. Characteristic parameters of physico-chemical transformations

A component is characterized by its chemical nature and by the phase to which it belongs. We can distinguish two types of physico-chemical transformations:

– physical transformations where there is no alteration of the chemical nature of the species but there is a modification of their phases;

– chemical transformations or reactions, where there is a modification of the chemical nature of the species present, and where that modification may or may not be accompanied by a phase change.

### 1.1.1. *Balance equation of a transformation*

Generally, a transformation is represented by its balance equation, which specifies the initial components, present before the transformation, and the end components. Symbolically, such a balance equation is written in the form:

$$0 = \sum_i \nu_i A_i \qquad [1R.1]$$

This formulation shows the algebraic stoichiometric numbers $\nu_i$, which are counted as positive for the end components and as negative for the initial components. The sum is extended to all of the initial and final components.

Physical transformations are characterized by the unit value of all the stoichiometric numbers.

It is often useful for the balance equation to show the nature of the phases in which the components are. In order to do this, a system of symbols as outlined in Table 1.1 is included in the balance equations. According to this system, the vaporization of pure liquid water in an atmosphere composed solely of water is written as:

$$(H_2O)=\{H_2O\} \qquad [1R.2]$$

| Symbol | Meaning | Symbol | Meaning |
|---|---|---|---|
| {A} | Pure A in a gaseous phase | {{A}} | A in a gaseous mixture |
| (A) | Pure A in a liquid phase | ((A)) | A in a liquid solution |
| <A> | Pure A in a solid phase | <<A>> | A in a solid solution |

**Table 1.1.** *Symbolic representation of the phases of components in balance equations*

The vaporization of water in air on the basis of a liquid solution is written as:

$$((H_2O))=\{\{H_2O\}\} \qquad [1R.3]$$

### 1.1.2. *Values associated with a transformation*

To each transformation, for any extensive value *A*, we attribute a value, written as $\Delta_r A$, called *A* associated with the transformation *r*. This value is defined, on the basis of the partial molar values $\overline{A_i}$ of the components involved in the transformation, by the relation:

$$\Delta_r A=\sum_i \nu_i \overline{A_i} \qquad [1.1]$$

Thus, we speak of the enthalpy, Gibbs energy, entropy, variations in volume, etc. associated with the transformation.

### 1.1.3. *Standard values associated with a transformation*

We shall see that in the particular case where we choose pure components in their phases, at a pressure of 1 bar, the *standard value A associated with the reaction* [1R.1] is defined by the relation:

$$\Delta_r A^0 = \sum_i \nu_i a_i^o \tag{1.2}$$

$a_i^o$ is the *standard value of the parameter A of the i*, meaning that the value of $A$ taken in the pure state at a pressure of 1 bar at the temperature of 298 K (25°C), the substance being in its normal state of condensation at that temperature and pressure. We thus define a *standard value at temperature T* as the above standard value chosen at the temperature $T$ instead of 298 K.

### 1.1.4. *Extent and rate of a transformation*

A transformation is also characterized by its extent $\xi$ at a time $t$ and its rate $\Re$ at that instant. The rate is the derivative in relation to time of the extent. If $n_k$ is the number of moles of the component $A_k$, we have:

$$\frac{d\xi}{dt} = \frac{1}{\nu_k}\frac{dn_k}{dt} \tag{1.3a}$$

and:

$$\Re = \frac{d\xi}{dt} \tag{1.3b}$$

In the case where multiple transformations take place in the system, equations [1.3a] and [1.3b] are replaced by:

$$\frac{dn_{k_\rho}}{dt} = \nu_{k_\rho}\Re_\rho \tag{1.4a}$$

and:

$$\frac{\mathrm{d}n_k}{\mathrm{d}t} = \sum_{\rho} \nu_{k_\rho} \Re_\rho \qquad [1.4b]$$

In these expressions, $\mathrm{d}n_{k_\rho}$, $\Re_\rho$ and $\nu_{k_\rho}$ respectively denote the variation of the quantity of the component $A_k$ due to the transformation $\rho$, the rate of the transformation $\rho$ and the stoichiometric number relative to component $A_k$ in the transformation $\rho$. The sum of relation [1.4b] is found for all of the transformations taking place in the system under study.

## 1.2. Entropy production during the course of a transformation in a closed system

Consider the transformation [1R.1]. We respectively denote by $\mathrm{d}_i S$ and $\mathrm{d}_e S$ the contributions to the entropy variation made by the entropy production within the system and the exchanges with the external medium. The entropy balance at each moment can be written as:

$$\frac{\mathrm{d}S}{\mathrm{d}t} = \frac{\mathrm{d}_i S}{\mathrm{d}t} + \frac{\mathrm{d}_e S}{\mathrm{d}t} \qquad [1.5]$$

For our study, we choose as variables the $p$ pertinent intensive variables $Y_k$ and the quantities of material whose fluxes are reduced, in a closed system, according to relation [1.3], to the derivative of the extent of the reaction. We can therefore express the entropy flux on the basis of those variables, so:

$$\frac{\mathrm{d}S}{\mathrm{d}t} = \sum_{k=1}^{p} \frac{\partial S}{\partial Y_k}\frac{\mathrm{d}Y_k}{\mathrm{d}t} + \frac{\partial S}{\partial \xi}\frac{\mathrm{d}\xi}{\mathrm{d}t} \qquad [1.6]$$

In addition, the entropy contribution due to the exchanges with the external environment is linked to the exchanged heat, by:

$$\frac{\mathrm{d}_e S}{\mathrm{d}t} = \frac{1}{T}\frac{\mathrm{d}Q}{\mathrm{d}t} \qquad [1.7]$$

This exchanged heat is expressed as a function of the variables by way of the specific heat coefficients $\chi_k$, which enables us to write the following for the entropy flux exchanged:

$$\frac{\mathrm{d}_e S}{\mathrm{d}t} = \sum_{k=1}^{p} \frac{\chi_k}{T} \frac{\mathrm{d}Y_k}{\mathrm{d}t} + \frac{\chi_T}{T} \frac{\mathrm{d}\xi}{\mathrm{d}t} \qquad [1.8]$$

By comparing expressions [1.6] and [1.8], we obtain the contribution of the internal production to the variation in entropy as a function of the system variables:

$$\frac{\mathrm{d}_i S}{\mathrm{d}t} = \sum_{k=1}^{p} \left( \frac{\partial S}{\partial Y_k} - \frac{\chi_k}{T} \right) \frac{\mathrm{d}Y_k}{\mathrm{d}t} + \left( \frac{\partial S}{\partial \xi} - \frac{\chi_T}{T} \right) \frac{\mathrm{d}\xi}{\mathrm{d}t} \qquad [1.9]$$

According to the second law, this entropy production must be positive or null in any spontaneous transformation. If we envisage a transformation whereby the external intensive variables $Y_k$ are kept constant, if the transformation is spontaneous, it means that we satisfy the inequality:

$$\frac{\mathrm{d}_i S}{\mathrm{d}t} = \left( \frac{\partial S}{\partial \xi} - \frac{\chi_T}{T} \right) \frac{\mathrm{d}\xi}{\mathrm{d}t} \geq 0 \qquad [1.10]$$

Thus, expression [1.10] is a condition needing to be fulfilled during any real transformation keeping the intensive variables $Y_k$ constant.

## 1.3. Affinity of a transformation

We shall introduce a new value – the affinity – pertaining to any transformation. The variables for this affinity are the (intensive or extensive) Thermodynamic variables, the quantities of material and the extent of the transformation.

### 1.3.1. *Definition*

De Donder proposed to use the term *affinity* of the transformation, denoted as $\mathscr{A}$, for the entity:

$$\mathscr{A} = T \left( \frac{\partial S}{\partial \xi} - \frac{\chi_T}{T} \right) \qquad [1.11]$$

According to expression [1.10], the entropy production can be expressed on the basis of that activity by:

$$\frac{\mathrm{d}_i S}{\mathrm{d}t} = \frac{\mathscr{A}}{T}\frac{\mathrm{d}\xi}{\mathrm{d}t} \qquad [1.12]$$

On the basis of this definition, we shall be able to express the affinity another way.

### 1.3.2. *Affinity and characteristic functions*

The variation of internal energy is, according to the first law of thermodynamics:

$$\frac{\mathrm{d}U}{\mathrm{d}t} = \frac{\mathrm{d}Q}{\mathrm{d}t} + \frac{\mathrm{d}W}{\mathrm{d}t} = T\frac{\mathrm{d}_e S}{\mathrm{d}t} + \frac{\mathrm{d}W}{\mathrm{d}t} = T\frac{\mathrm{d}S}{\mathrm{d}t} - T\frac{\mathrm{d}_i S}{\mathrm{d}t} + \frac{\mathrm{d}W}{\mathrm{d}t} \qquad [1.13]$$

If all the exchanges are reversible, apart from the transformation under study (we then say that the system is at physical equilibrium), then the work term is written:

$$\frac{\mathrm{d}W}{\mathrm{d}t} = \sum_{k=2}^{p} Y_k dX_k \qquad [1.14]$$

The sum that appears in the above expression is extended to all couples of conjugate variables, with the exception of the temperature–entropy couple (which is why the index *k* begins at the value of 2). By substituting this back into expression [1.13] and taking account of relation [1.12], we obtain:

$$\frac{\mathrm{d}U}{\mathrm{d}t} = \sum_{k=2}^{p} Y_k \frac{\mathrm{d}X_k}{\mathrm{d}t} + T\frac{\mathrm{d}S}{\mathrm{d}t} - \mathscr{A}\frac{\mathrm{d}\xi}{\mathrm{d}t} \qquad [1.15]$$

This gives us a new expression of the affinity, which is therefore the opposite of the differential of the function $U$ in relation to the fractional extent with constant entropy and extensive variables:

$$\mathscr{A} = -\left(\frac{\partial U}{\partial \xi}\right)_{S,X_k} \qquad [1.16]$$

We can generalize this expression for any characteristic function $\Gamma$, defined by:

$$\Gamma = U - TS - \sum_{i=2}^{q} X_i Y_i \text{ where } 2 \leq q \leq p \qquad [1.17]$$

By differentiation of [1.17], we obtain:

$$\mathrm{d}\Gamma = \mathrm{d}U - T\,\mathrm{d}S - S\,\mathrm{d}T - \sum_{i=2}^{q} X_i\,\mathrm{d}Y_i - \sum_{i=2}^{q} Y_i dX_i \qquad [1.18]$$

Thus, when we consider relation [1.15]:

$$\mathrm{d}\Gamma = -S\,\mathrm{d}T - \sum_{i=2}^{q} X_i\,\mathrm{d}Y_i + \sum_{m=q+1}^{p} Y_m dX_m - \mathscr{A}\,\mathrm{d}\xi \qquad [1.19]$$

we obtain a new expression of the affinity which generalizes relation [1.16]:

$$\mathscr{A} = -\left(\frac{\partial \Gamma}{\partial \xi}\right)_{X_m Y_l T} \qquad [1.20]$$

In particular, for chemical systems with the variables pressure and temperature, the characteristic function is the Gibbs energy $G$. We obtain:

$$\mathscr{A} = -\left(\frac{\partial G}{\partial \xi}\right)_{P,T} \qquad [1.21]$$

At constant pressure and temperature, the affinity is the opposite of the partial derivative of the Gibbs energy in relation to the extent.

NOTE 1.1.– Expression [1.15] shows that the affinity, which is an extensive value, and the extent, which is an intensive value, are two conjugate values.

### 1.3.3. *Affinity and chemical potentials*

If we consider relation [1.3], we can write:

$$\left(\frac{\partial G}{\partial \xi}\right)_{P,T} = \sum_i \nu_i \left(\frac{\partial G}{\partial n_i}\right)_{P,T,n_j} = \sum_i \nu_i \mu_i \qquad [1.22]$$

By substituting this back into relation [1.21], we obtain an expression of the affinity as a function of the chemical potentials of the components involved in the transformation:

$$\mathscr{A} = -\sum_i \nu_i \mu_i \qquad [1.23]$$

The affinity of a transformation therefore depends only on the chemical potentials of the components involved in that transformation.

Using expression [1.22], we can write:

$$\sum_i \nu_i \left(\frac{\partial G}{\partial n_i}\right)_{P,T,n_j} = \sum_i \nu_i \overline{G}_i = \Delta_r G = \sum_i \nu_i \mu_i \qquad [1.24]$$

Thus, by comparing this with relation [1.23]:

$$\Delta_r G = -\mathscr{A} \qquad [1.25]$$

The affinity of a transformation is thus the opposite of the Gibbs energy associated with that transformation.

These results can easily be generalized to any general Gibbs energy using the generalized chemical potentials which correspond to it. For example, for the electrochemical Gibbs energy and the electrochemical potentials, an expression such as [1.23] will give the electrochemical affinity of an electrochemical reaction.

### 1.3.4. *Affinity, reaction quotient and activities*

If, in relation [1.23], we explicitly state the chemical potentials of the species in solution in the form:

$$\mu_i = \mu_i^0 + \mathrm{R}T \ln a_i \qquad [1.26]$$

for the transformation [1R.1], we find the expression of the affinity as a function of the activities:

$$\mathscr{A} = -\sum_i \nu_i \mu_i^0 - \mathrm{R}T \prod_i a_i^{\nu_i} \qquad [1.27]$$

If we define the value $Q_r$, known as the *reaction quotient* of the transformation $r$, by the relation:

$$Q_r = \prod_i a_i^{\nu_i} \qquad [1.28]$$

then the affinity takes the form:

$$\mathscr{A} = \mathscr{A}^0 - RT \ln Q_r \qquad [1.29]$$

Using relation [1.25], the Gibbs energy associated with the reaction takes the form:

$$\Delta_r G = \Delta_r G^0 + RT \ln Q_r \qquad [1.30]$$

The last two relations will be useful for the expression of the equilibrium constants (see section 3.1).

### 1.3.5. *Total differential of the affinity in variables $Y_l$, $X_m$, $\xi$*

As the affinity is a function of state, its differential, expressed on the basis of the chosen variables, will be of the form:

$$\mathrm{d}A = \sum_l \frac{\partial \mathscr{A}}{\partial Y_l} \mathrm{d}Y_l + \sum_m \frac{\partial \mathscr{A}}{\partial X_m} \mathrm{d}X_m + \frac{\partial \mathscr{A}}{\partial \xi} \mathrm{d}\xi \qquad [1.31]$$

Using relation [1.20], we find:

$$\frac{\partial \mathscr{A}}{\partial \xi} = -\frac{\partial^2 \Gamma}{\partial \xi^2} \qquad [1.32]$$

and by applying equation [1.23]:

$$\frac{\partial \mathscr{A}}{\partial Z} = -\sum_i \nu_i \frac{\partial \mu_i}{\partial Z} \qquad [1.33]$$

with $Z$ being one of the variables in the set ($Y_l$,$X_m$).

In the case where that set of variables comprises only intensive variables $Y_l$, the differential of the affinity is written:

$$\mathrm{d}A = \sum_l \frac{\partial \mathcal{A}}{\partial Y_l}\mathrm{d}Y_l + \frac{\partial \mathcal{A}}{\partial \xi}\mathrm{d}\xi \qquad [1.34]$$

However, by applying relation [1.33] and the symmetry of the characteristic matrix, if $X_l$ is the conjugate extensive value of $Y_l$ , we obtain:

$$\frac{\partial A}{\partial Y_l} = \sum_i \nu_i \overline{X}_{l_{i_i}} = \Delta_r X_l \qquad [1.35]$$

This gives us the differential of the affinity in intensive variables (partial molar values) and extent:

$$\mathrm{d}\mathcal{A} = \Delta_r X_l \,\mathrm{d}Y_l + \frac{\partial \mathcal{A}}{\partial \xi}\mathrm{d}\xi \qquad [1.36]$$

Let us apply these results to chemical systems with variables $-P$, $T$, so then equation [1.35] gives us:

$$\frac{\partial \mathcal{A}}{\partial T} = \sum_i \nu_i \overline{S}_{i_i} = \Delta_r S \qquad [1.37]$$

In addition, we have:

$$\frac{\partial \mathcal{A}}{\partial P} = -\sum_i \nu_i \overline{V}_{i_i} = -\Delta_r V \qquad [1.38]$$

The differential of the affinity then becomes:

$$\mathrm{d}\mathcal{A} = \Delta_r S \,\mathrm{d}T - \Delta_r V \,\mathrm{d}P - \frac{\partial^2 G}{\partial \xi^2}\mathrm{d}\xi \qquad [1.39]$$

NOTE 1.2.– Helmholtz's second relation gives the derivative of the ratio $\mu_i/T$ with the temperature:

$$\left[\frac{\partial\left(\frac{\mu_i}{T}\right)}{\partial T}\right]_{P,\xi} = \frac{\overline{H}_{i_i}}{T^2} \qquad [1.40]$$

By coupling this relation with expression [1.23], we obtain:

$$\left[\frac{\partial\left(\frac{\mathscr{A}}{T}\right)}{\partial T}\right]_{P,\xi} = \frac{\sum\limits_i \nu_i \overline{H}_{i_i}}{T^2} = \frac{\Delta_r H}{T^2} \qquad [1.41]$$

$\Delta_r H$ is the enthalpy associated with the transformation studied.

### 1.3.6. *Derivatives of the affinity in relation to the extent and the chemical potentials*

By deriving equation [1.23], we obtain:

$$\frac{\partial \mathscr{A}}{\partial \xi} = -\sum_i \nu_i \frac{\partial \mu_i}{\partial \xi} \qquad [1.42]$$

However, by taking account of relation [1.3], we can write:

$$\frac{\partial \mu_i}{\partial \xi} = \sum_{k=1}^{N} \frac{\partial \mu_k}{\partial n_k}\frac{\partial n_k}{\partial \xi} = \sum_k \nu_k \frac{\partial \mu_i}{\partial n_k} \qquad [1.43]$$

Thus:

$$\frac{\partial \mathscr{A}}{\partial \xi} = -\sum_k \sum_i \nu_k \nu_i \frac{\partial \mu_i}{\partial n_k} \qquad [1.44]$$

which can be expressed in the form:

$$\frac{\partial \mathscr{A}}{\partial \xi} = -\sum_k \sum_{k \neq i} \nu_k \nu_i \frac{\partial \mu_i}{\partial n_k} + \sum_i \nu_i^2 \frac{n_i^2}{n_i^2}\frac{\partial \mu_i}{\partial n_i} \qquad [1.45]$$

However, according to the Gibbs-Duhem relation, we have:

$$n_i \frac{\partial \mu_i}{\partial n_i} = -\sum_{k \neq i} n_k \frac{\partial \mu_k}{\partial n_i} \qquad [1.46a]$$

The symmetry of the characteristic matrix leads to:

$$\frac{\partial \mu_i}{\partial n_k} = \frac{\partial \mu_k}{\partial n_i} \tag{1.46b}$$

Thus, by substituting the expressions [1.46a] and [1.46b] into expression [1.45], we find:

$$\frac{\partial \mathscr{A}}{\partial \xi} = -\sum_k \sum_{k \neq i} \nu_k \nu_i \frac{\partial \mu_i}{\partial n_k} + \sum_i \sum_{k \neq i} \nu_i^2 \frac{n_i n_k}{n_i^2} \frac{\partial \mu_i}{\partial n_i} \tag{1.47}$$

By decomposing the second term, this expression takes the form:

$$\frac{\partial \mathscr{A}}{\partial \xi} = -\sum_k \sum_{k \neq i} \nu_k \nu_i \frac{\partial \mu_i}{\partial n_k} - \frac{1}{2} \sum_i \sum_{k \neq i} \nu_i^2 \frac{n_i^2}{n_i^2} \frac{\partial \mu_i}{\partial n_i} + \frac{1}{2} \sum_i \sum_{k \neq i} \nu_k^2 \frac{n_i^2}{n_i^2} \frac{\partial \mu_i}{\partial n_k}$$

We can verify that the above formula is equivalent to:

$$\frac{\partial \mathscr{A}}{\partial \xi} = \frac{1}{2} \sum_i \sum_k \frac{\partial \mu_i}{\partial n_k} \left( \frac{\nu_i}{n_i} - \frac{\nu_k}{n_k} \right)^2 n_i n_k \tag{1.48}$$

This expression gives us the derivative of the affinity in relation to the extent as a function of the derivatives of the chemical potentials of the components of the reaction in relation to the quantities of the other materials.

## 1.4. De Donder's inequality – direction of the transformations and equilibrium conditions

The second principle, applied to relation [1R.1], gives us:

$$\frac{\mathscr{A}}{T} \frac{\mathrm{d}\xi}{\mathrm{d}t} \geq 0 \tag{1.49}$$

In this inequality, we see the appearance of the rate of reaction (expression [1.2]), which is tantamount to writing that in order for a transformation to be possible, it is necessary for the affinity and rate to obey the condition:

$$\mathscr{A}\Re \geq 0 \tag{1.50}$$

This inequality constitutes *de Donder's inequality*, which is a general formulation of the condition of transformation and does not depend on the set of variables chosen.

We shall now discuss this inequality:

– if the affinity is positive, we need to have a positive or null rate to respect the condition [1.50]. If the rate is positive, this means that the transformation takes place from left to right in the chosen formula. If the rate is zero, the transformation does not happen;

– if the affinity is negative, the inequality gives a negative or null rate, and the transformation will take place spontaneously from right to left of the chosen formula, or will not take place;

– if the affinity is null, then the rate is null because, if such were not the case, the system would be the seat of a reversible transformation (null entropy production) with a non-null rate, which is incompatible with the definition of a reversible transformation.

Let us now examine the reciprocal of our discussion:

– if the rate is positive then inequality [1.50] necessarily leads to a positive affinity. The affinity cannot be null because we would still have a reversible transformation at non-null rate;

– if the rate is negative, then the affinity is also negative, with the reverse reaction occurring;

– if the rate is null, then the system is at equilibrium. We can therefore have either a null or a non-null affinity. In the first case (null affinity), we say that the system is at *thermodynamic equilibrium*. In the second case (non-null affinity), the system is said to be at *false equilibrium*. False equilibrium is thus encountered for a transformation which does not take place although its affinity is positive. For instance, we can cite the synthesis of water under normal conditions of temperature and pressure. The hydrogen-oxygen mixture does not react. Its rate is too slow for the reaction to be perceptible. In such a case, we could use a catalyst to carry out such a transformation, which would not alter the affinity but would modify the rate. The choice of a catalyzer is useless and ineffectual if the affinity is null or negative.

Thus, the necessary and sufficient condition of thermodynamic equilibrium is, by definition:

$$\mathscr{A} = 0 \tag{1.51}$$

and the true meaning of the possibility of occurrence of the transformation is given by the condition:

$$\mathscr{A} > 0 \tag{1.52}$$

In light of relations [1.23] and [1.24], these conditions are also written:

For thermodynamic equilibrium:

$$\Delta_r G = 0 \tag{1.53}$$

or:

$$\sum_{k=1}^{N} \nu_k \mu_k = 0 \tag{1.54}$$

For the direction of spontaneous reaction:

$$\Delta_r G < 0 \tag{1.55}$$

or:

$$\sum_{k=1}^{N} \nu_k \mu_k < 0 \tag{1.56}$$

These conditions are independent of the nature of the physico-chemical transformation and the variables chosen to define the system, provided that system is home only to a single transformation.

## 1.5. Heats of transformation

Usually, transformations involve heat exchanges with the outside environment, which may be due to heat being released (exothermic transformations) or absorbed (in the case of endothermic transformations). We shall express those heat exchanges in two important cases.

### 1.5.1. *Heat of transformation at constant pressure and temperature*

We can express the elementary heat of a transformation in the form:

$$\delta Q = T\,\mathrm{d}_e S = T\,\mathrm{d}S - T\,\mathrm{d}_i S \qquad [1.57]$$

By choosing all the variables ($T$, $-P$, $\xi$), we shall have:

$$\delta Q = T\frac{\partial S}{\partial T}\mathrm{d}T + T\frac{\partial S}{\partial P}\mathrm{d}P + T\frac{\partial S}{\partial \xi}\mathrm{d}\xi - T\,\mathrm{d}_i S \qquad [1.58]$$

At constant pressure and temperature, this gives us:

$$\delta Q_P = T\frac{\partial S}{\partial \xi}\mathrm{d}\xi - T\,\mathrm{d}_i S \qquad [1.59]$$

However, as the generalized Gibbs energy $\Gamma$ is a function of state, we have, according to relation [1.19]:

$$\left(\frac{\partial S}{\partial \xi}\right)_{P,T} = -\left(\frac{\partial \mathscr{A}}{\partial T}\right)_{P,\xi} \qquad [1.60]$$

Thus, we can write relation [1.59] in the form:

$$\delta Q_P = -\left[T\left(\frac{\partial \mathscr{A}}{\partial T}\right)_{P,\xi} + \mathscr{A}\right]\mathrm{d}\xi \qquad [1.61]$$

By integrating this expression for the whole of the transformation, we obtain:

$$Q_P = \int_{\xi=0}^{\xi=\infty} \delta Q_P = -\left[T\left(\frac{\partial \mathscr{A}}{\partial T}\right)_{P,\xi} + \mathscr{A}\right] \qquad [1.62]$$

By replacing the affinity with its expression [1.22], we find:

$$Q_P = T\sum_k \nu_k \frac{\partial \mu_k}{\partial T} + \sum_k \nu_k \mu_k \qquad [1.63]$$

Thus, by using the symmetry of the characteristic matrix:

$$Q_P = T\sum_k \nu_k \overline{S}_k + \sum_k \nu_k \mu_k \qquad [1.64]$$

However, we know that we have:

$$\mu_k = \overline{G}_k = \overline{H}_k - T\overline{S}_k \qquad [1.65]$$

For the heat of reaction we find:

$$Q_P = \sum_k \nu_k \overline{H}_k = \Delta_r H \qquad [1.66]$$

Thus, the heat of transformation, at constant temperature and pressure, is equal to the enthalpy associated with the reaction.

NOTE 1.3.– For a long time, chemists counted the heat released by an exothermic reaction positively. With this old convention, relation [1.66] was transformed into:

$$Q_P = -\Delta_r H$$

### 1.5.2. *Heat of transformation at constant volume and temperature*

Let us look again at expression [1.58]. By choosing the set of variables ($T$, $V$, $\xi$), we obtain:

$$\delta Q = T\frac{\partial S}{\partial T}\mathrm{d}T + T\frac{\partial S}{\partial V}\mathrm{d}V + T\frac{\partial S}{\partial \xi}\mathrm{d}\xi - T\mathrm{d}_i S \qquad [1.67]$$

At constant volume and temperature we have:

$$\delta Q_V = T\frac{\partial S}{\partial \xi}\mathrm{d}\xi - T\mathrm{d}_i S \qquad [1.68]$$

By applying the same reasoning as in the previous section (see section 1.5.1), we find:

$$Q_V = T\sum_k \nu_k \left(\frac{\partial S}{\partial n_k}\right)_{V,T,n_j} + \sum_k \nu_k \mu_k \quad [1.69]$$

However, we know that:

$$\mu_i = \left(\frac{\partial F}{\partial n_k}\right)_{V,T,n_j} \quad [1.70]$$

Thus:

$$Q_V = T\sum_k \nu_k \left(\frac{\partial S}{\partial n_k}\right)_{V,T,n_j} + \sum_k \nu_k \left(\frac{\partial F}{\partial n_k}\right)_{V,T,n_j} \quad [1.71]$$

and, taking account of:

$$F = U - TS \quad [1.72]$$

the heat of transformation is:

$$Q_V = \sum_k \nu_k \left(\frac{\partial U}{\partial n_k}\right)_{V,T,n_j} = \Delta_r U \quad [1.73]$$

Thus, the heat of transformation, at constant temperature and volume, is equal to the internal energy associated with the reaction.

### 1.5.3. *Variations in the heat of transformation at constant pressure with changing temperature – Kirchhoff relation*

According to equation [1.66], we can write:

$$\left(\frac{\partial Q_P}{\partial T}\right)_P = \sum_k \nu_k \left(\frac{\partial \overline{H}_k}{\partial T}\right)_{P_j} \quad [1.74]$$

However, the differential of the partial molar enthalpy can be expressed in the form:

$$\mathrm{d}\overline{H}_k = T\,\mathrm{d}S_k + \overline{V}_k\,\mathrm{d}P + \sum_k \mu_k\,\mathrm{d}n_k \qquad [1.75]$$

Thus, at a given pressure and extent, we obtain:

$$\left(\frac{\partial \overline{H}_k}{\partial T}\right)_{P_j} = T\left(\frac{\partial \overline{S}_k}{\partial T}\right)_{P_j} = \overline{C}_{P_k} \qquad [1.76]$$

By substituting that equation back into relation [1.74], we find:

$$\left(\frac{\partial Q_P}{\partial T}\right)_P = \sum_k \nu_k \overline{C}_{P_k} = \Delta_r C_P \qquad [1.77]$$

This relation constitutes what we call the *Kirchhoff relation.* An equivalent relation would give the variation of the heat of transformation at constant volume with the temperature as a function of the molar specific heat capacity at constant volume associated with the transformation.

## 1.6. Set of points representing the equilibrium states of a transformation

Consider a transformation at thermodynamic equilibrium for given values of the variables of state. Its affinity is null. If we vary one or more of the variables of state by an infinitesimal amount, the affinity takes on a new value $\mathscr{A} + \mathrm{d}\mathscr{A}$. In order for that new state to also be a state of equilibrium of transformation, it is necessary for that new value of the affinity to be null, so:

$$\mathscr{A} + \mathrm{d}\mathscr{A} = 0 \qquad [1.78]$$

This leads us to:

$$\mathrm{d}\mathscr{A} = 0 \qquad [1.79]$$

By stating the affinity on the basis of the chemical potentials of the $N$ components involved in the transformation, the condition becomes:

$$\sum_{k=1}^{N} \nu_k \, \mathrm{d}\mu_k = 0 \qquad [1.80]$$

Either of these last two expressions, [1.79] and [1.80], enables us to define the set of equilibrium states of the transformation.

## 1.7. Closed systems accommodating multiple reactions

Consider a system in which, between the components, there are $R$ possible transformations. A component may not necessarily be involved in multiple kinds of transformations (in a transformation in which it is not involved, its stoichiometric coefficient will be zero). For the $\rho^{\text{th}}$ transformation, we can define an affinity according to relation [1.11], and by applying relation [1.23], we find:

$$\mathscr{A}_\rho = -\sum_k \nu_{k_\rho} \mu_k \qquad [1.81]$$

If $\nu_{k_\rho}$ is the stoichiometric number of component $A_k$ in the $\rho^{\text{th}}$ transformation, then by pursuing the same reasoning as in section 1.3.2, with $\xi_\rho$ representing the extent of the reaction $\rho$, we obtain:

$$\mathscr{A}_\rho = -\frac{\partial \Gamma}{\partial \xi_\rho} \qquad [1.82]$$

The real transformation condition, therefore, will be:

$$\mathrm{d}\Gamma = -\sum_\rho \mathscr{A}_\rho \, \mathrm{d}\xi_\rho \le 0 \qquad [1.83]$$

and consequently:

$$\sum_\rho \mathscr{A}_\rho \frac{\mathrm{d}\xi_\rho}{\mathrm{d}t} = \sum_\rho \mathscr{A}_\rho \Re_\rho \ge 0 \qquad [1.84]$$

This inequality is a generalization of De Donder's inequality (see section 1.4). We can see that, for a transformation to be possible, it is no longer necessary for inequality [1.50] to be satisfied, if that transformation occurs in a system which contains multiple transformations, and such that relation [1.84] will, itself, be satisfied. This phenomenon of a reaction which is impossible on its own but is possible within a set of transformations, in the same conditions, is known as *chemical coupling*.

At thermodynamic equilibrium, the sum appearing in relation [1.83] must be zero, implying that all the individual terms $\mathscr{A}_\rho$ are zero regardless of the transformation at hand. The condition of equilibrium of transformations in the system, for any transformation $\rho$ belonging to the set $R$, will therefore be:

$$\mathscr{A}_\rho = 0 \qquad [1.85]$$

NOTE 1.4.– We shall see (in section 2.2) that this condition is sufficient but not necessary if not all the transformations are independent.

## 1.8. Direction of evolution and equilibrium conditions in an open system

In an open system, the variations $\mathrm{d}n_k$ in the amount of component $A_k$ are no longer linked to one another by the transformation; these quantities can also change because of exchanges of matter with the external environment. In a chemical system, the variation of Gibbs energy is written in the form:

$$\mathrm{d}G = -S\mathrm{d}T + V\mathrm{d}P + \sum_{k=1}^{N} \mu_k \,\mathrm{d}n_k \qquad [1.86]$$

For equilibria at temperature and pressure that are both kept constant, the Gibbs energy is a potential function and therefore in order for the system to be able to evolve spontaneously, we need:

$$\mathrm{d}G = \sum_{k=1}^{N} \mu_k \,\mathrm{d}n_k \leq 0 \qquad [1.87]$$

At thermodynamic equilibrium, the potential function reaches an extremum. If that extremum is a minimum, then the equilibrium is stable. If the extremum is a maximum, then the equilibrium is unstable (see section 2.6.2).

At equilibrium, we therefore ought to have:

$$\sum_{k=1}^{N} \mu_k \, \mathrm{d} n_k = 0 \qquad [1.88]$$

This relation is a generalization of relation [1.54], which takes that form when we remember that, in a closed environment where only a single transformation takes place, by definition we have the following formula for the extent:

$$\mathrm{d} n_k = \nu_k \, \mathrm{d} \xi \qquad [1.89]$$

## 1.9. Azeotropic transformations

A closed system undergoes an azeotropic transformation when, during the course of the transformation, the masses of some of the phases increase at the expense of others, without a change in the composition of the phases. This is expressed, for all phases α and all compositions, by the property:

$$x_k^{(\alpha)} = x_k^{0(\alpha)} \qquad [1.90]$$

$x_k^{0(\alpha)}$ denotes the molar fraction of component $A_k$ in phase α at the initial time of the transformation and $x_k^{(\alpha)}$ its molar fraction at any given moment during the transformation.

By deriving equation [1.90] in relation to time, we find the following for any component $A_k$ in any phase α:

$$\frac{\mathrm{d} x_k^{(\alpha)}}{\mathrm{d} t} = 0 \qquad [1.91]$$

In view of the definition of the molar fractions, this relation is written as:

$$\frac{\mathrm{d} n_k^{(\alpha)}}{\mathrm{d} t} - x_k^{0(\alpha)} \sum_{i=1}^{N} \frac{\mathrm{d} n_i^{(\alpha)}}{\mathrm{d} t} = 0 \qquad [1.92]$$

We can apply this condition to a system with only one transformation with the fractional extent $\xi$. We then obtain:

$$\frac{\mathrm{d}\xi}{\mathrm{d}t} = x_k^{0(\alpha)} \frac{\mathrm{d}\xi}{\mathrm{d}t} \sum_{i=1}^{N} \nu_i^{(\alpha)} = 0 \qquad [1.93]$$

or:

$$x_k^{0(\alpha)} = \frac{\nu_k^{(\alpha)}}{\sum_{i=1}^{N} \nu_i^{\alpha}} \qquad [1.94]$$

At the initial time, all of the components must be in stoichiometric proportions in each phase.

Another case encountered when we look at phase-change in multi-component systems is when each component is involved in only one transformation, and its stoichiometric number is 1. Thus, in a phase α, we have:

$$\frac{\mathrm{d}\xi_k^{(\alpha)}}{\mathrm{d}t} = x_k^{0(\alpha)} \sum_{i=1}^{N} \nu_i \frac{\mathrm{d}\xi_i^{(\alpha)}}{\mathrm{d}t} \qquad [1.95]$$

This can also be written as:

$$\frac{\frac{\mathrm{d}\xi_i^{(\alpha)}}{\mathrm{d}t}}{\frac{\mathrm{d}\xi_k^{(\alpha)}}{\mathrm{d}t}} = \frac{x_i^{0(\alpha)}}{x_k^{0(\alpha)}} \qquad [1.96]$$

This means that the transformation rates of two components are in a constant ratio to one another, with the value of that ratio being determined by the initial conditions.

In addition, if we consider two phases α and β, by applying relation [1.92] for component $A_k$ in the two phases and adding together the expressions obtained, we find:

$$\frac{\mathrm{d}n_k^{(\alpha)}}{\mathrm{d}t} + \frac{\mathrm{d}n_k^{(\beta)}}{\mathrm{d}t} - x_k^{0(\alpha)} \sum_{i=1}^{N} \frac{\mathrm{d}n_i^{(\alpha)}}{\mathrm{d}t} - x_k^{0(\beta)} \sum_{i=1}^{N} \frac{\mathrm{d}n_i^{(\beta)}}{\mathrm{d}t} = 0 \qquad [1.97]$$

However, as we are dealing with the case where there is only one transformation per component, we necessarily have stoichiometric numbers of 1:

$$\frac{dn_k^{(\alpha)}}{dt} = -\frac{dn_k^{(\beta)}}{dt} \qquad [1.98]$$

and:

$$\sum_{i=1}^{N} \frac{dn_i^{(\alpha)}}{dt} = -\sum_{i=1}^{N} \frac{dn_i^{(\beta)}}{dt} \qquad [1.99]$$

Finally, we deduce from this:

$$x_k^{0(\alpha)} = x_k^{0(\beta)} \qquad [1.100]$$

In the initial state, the molar fractions are the same in both phases.

Thus, the necessary and sufficient conditions for a phase change to be an azeotropic transformation are that:

– the system starts in an equi-content initial state;

– the ratio of the transfer rates is constant over time.

NOTE 1.5.– The azeotropic nature of the transformation pertains only to the compositions of the phases; it is independent of the external intensive variables (temperature, pressure, etc.) insofar as the azeotropic nature of the process only covers the compositions of phases; it is not dependent on external intensive variables (temperature, pressure, etc.), because all the kinetic laws of transition from one phase to another are identical functions of these variables.

The results in this section never entail the hypothesis that the transformations are at equilibrium; they are just as applicable for true transformations as for reversible transformations. In the latter case, the rates are null, and we are left with the condition of equi-content [1.100].

An example of azeotropic transformations, besides certain phase changes, includes the transformation:

$$\{\{HCl\}\} + \{\{NH_3\}\} = < NH_4Cl > \qquad [1R.4]$$

If we start with the three components, the gaseous phase is equimolecular in ammonia and hydrogen chloride.

Another example is given by the decomposition of calcium carbonate:

$$< \mathrm{Ca\,CO_3} > = < \mathrm{CaO} > + \{\mathrm{CO_2}\} \qquad [1R.5]$$

If, at the initial time, we begin with a mixture of the three components in arbitrary proportions because they belong to different phases, the transformation is always azeotropic.

# 2

# Properties of States of Physico-Chemical Equilibrium

Having set out the conditions of equilibrium of a transformation in Chapter 1, in this chapter we shall examine certain properties of those states of equilibrium.

## 2.1. Laws of displacement of an equilibrium

Consider a system in which a transformation takes place which we suppose to be at thermodynamic equilibrium at certain variable values. We propose to examine the state to which the system will evolve if we alter the values of some of those variables. We shall suppose, for the time being, that the equilibrium is not broken – i.e. that the new state is also an equilibrium state. The evolution after equilibrium rupture will be examined later on (see section 2.6)

### 2.1.1. *General form of the displacement laws*

As the system is initially at equilibrium, the affinity of the transformation is null. As this affinity is a state function, if we alter one or more of these variables even by an infinitesimal amount, the affinity varies by $d\mathscr{A}$, and becomes:

$$\mathscr{A}' = d\mathscr{A}$$

If we apply de Donder's inequality to the new state of rate $\Re'$, we obtain:

$$\mathrm{d}\mathscr{A}\Re' \geq 0 \tag{2.1}$$

This relation gives us the general form of the system's evolution after the disturbance. In terms of chemical potentials, this inequality becomes:

$$\sum_k \nu_k d\mu_k \Re' < 0 \tag{2.2}$$

We shall examine a few particular cases.

### 2.1.2. *Influence of a temperature disturbance*

In terms of the variables pressure, temperature, and compositions, the affinity is expressed by relation [1.39]. If we apply a disturbance to the system at equilibrium by varying only the temperature – i.e. if the pressure and extent of the reaction remain constant – the new value of the affinity becomes:

$$\mathrm{d}\mathscr{A} = \Delta_r S \,\mathrm{d}T \tag{2.3}$$

However, at equilibrium, we have:

$$\Delta_r G = \Delta_r H - T\,\Delta_r S = 0 \tag{2.4}$$

Consequently, by extracting the entropy from that equation and substituting it back into expression [2.3], we obtain:

$$\mathrm{d}\mathscr{A} = \frac{\Delta_r H}{T}\mathrm{d}T \tag{2.5}$$

Suppose that we have an elevation in temperature ($\mathrm{d}T > 0$). By virtue of relations [2.1] and [2.5], we would have:

– if $\Delta_r H > 0$ $\mathrm{d}\mathscr{A} > 0$ and therefore $\Re' > 0$

– if $\Delta_r H < 0$, $\mathrm{d}\mathscr{A} < 0$ and therefore $\Re' < 0$

Hence, a temperature increase shifts the equilibrium in the endothermic direction ($\Delta_r H > 0$) of the transformation.

### 2.1.3. *Influence of a pressure disturbance*

Now, the temperature and the extent of the reaction are kept constant, and we provoke a disturbance of the equilibrium by a slight variation in pressure, which causes a variation of the affinity which is, by virtue of relation [1.39]:

$$\mathrm{d}\mathsf{A} = -\Delta_r V \,\mathrm{d}P \qquad [2.6]$$

If we choose to increase pressure ($\mathrm{d}P > 0$), it causes a decrease in volume. Because the variation of the affinity has the opposite sign to that of the volume, according to relation [2.6], we obtain:

– if $\Delta_r V < 0$, $\mathrm{d}\mathcal{A} > 0$ and therefore $\Re' > 0$;

– if $\Delta_r V > 0$, $\mathrm{d}\mathcal{A} < 0$ and therefore $\Re' < 0$.

Thus, an increase in pressure will favor the displacement of the equilibrium toward a new state of equilibrium, in the direction which is accompanied by a negative associated volume of the reaction ($\Delta_r V < 0$). If a gas is present amongst the reagents and/or reaction products, the number of gaseous moles decrease. In the case of condensed phases alone, the pressure will have practically no influence.

### 2.1.4. *Influence of the addition of a component*

We now keep the temperature and pressure constant, as well as the quantities of all the different components of a system except for one, because the system is disturbed by adding $\mathrm{d}n_i$ moles of component $A_i$. The affinity of the disturbed transformation will, in light of relation [1.23], be:

$$\mathrm{d}\mathcal{A} = -\sum_k \nu_k \frac{\partial \mu_k}{\partial n_i} \mathrm{d}n_i \qquad [2.7]$$

By separating the contribution of the added component $A_i$, this expression can also be written as:

$$\mathrm{d}\mathcal{A} = -\sum_{k \neq i} \nu_k \frac{\partial \mu_k}{\partial n_i} \mathrm{d}n_i - \nu_i \frac{\partial \mu_i}{\partial n_i} \mathrm{d}n_i \qquad [2.8]$$

To proceed further, we need to know the variations of the chemical potentials of the different components and the quantities of those components. Two cases need to be considered. In the first case, component $A_i$ is pure in its condensed phase. In the second case, component $A_i$ is either a pure gas or it belongs to a polycomponent phase of the system.

If the component added is pure in a condensed phase, the chemical potential of the other components, which are absent in the same phase will be: $\partial\mu_k / \partial n_i = 0$. The chemical potential of the added component, which is only present in the phase will be: $\partial\mu_i / \partial n_i = 0$. The addition of component $A_i$ does not alter the state of equilibrium. This case is encountered, for example, when we add any one of the solid components (carbonate or lime) into the decomposition of calcium carbonate by reaction [1R.5].

If the component added is gaseous or belongs to one of the polycomponent phases of the system, the application of relation [2.8] to inequality [2.1] depends on the variations of the chemical potentials with the quantity of the added component, i.e. definitively of the law of variation of the activities of the components belonging to the same phase as that which has been added. We know that there is no general law. The laws issuing from the different models of solution can be applied, but the result will depend on the model chosen.

To illustrate the displacement of equilibrium by adding a certain amount of a component, consider a perfect solution, which will immediately illustrate the diversity of the results obtained. We assign the relative values to the component added with index *i*; for the other components we use index *k*. When all the components are involved, we use the index *j* ($j = i + k$).

In the case of a perfect solution, the expressions of the chemical potential and of the molar fraction, for variations of the chemical potentials of the existing components ($k \neq i$), yield the relation:

$$\frac{\partial\mu_k}{\partial n_i} = -\frac{\mathrm{R}T}{\sum_j n_j} \qquad [2.9]$$

Similarly, for the chemical potential of the component $i$, which is added, we obtain:

$$\frac{\partial \mu_i}{\partial n_i} = \frac{RT}{\sum_j n_j} \left( \frac{\sum_j n_j}{n_i} - 1 \right) \qquad [2.10]$$

By substituting expressions [2.9] and [2.10] back into equation [2.8], we find:

$$\frac{\partial \mathscr{A}}{\partial n_i} = \frac{RT}{\sum_j n_j} \left( \sum_j \nu_j - \nu_i \frac{\sum_j n_j}{n_i} \right) \qquad [2.11]$$

and the direction of evolution is obtained by feeding expression [2.11] back into equation [2.1], which gives us:

$$\frac{RT}{\sum_j n_j} \left( \sum_j \nu_j - \nu_i \frac{\sum_j n_j}{n_i} \right) \Re' > 0 \qquad [2.12]$$

This inequality can be written more simply, using $x_i$ to denote the molar fraction of the component added in its phase, in the form:

$$\left( \sum_j \nu_j - \frac{\nu_i}{x_i} \right) \Re' > 0 \qquad [2.13]$$

In order to draw conclusions, we need to discuss inequality [2.13].

*First case*: the transformation takes place without varying the number of moles in the phase of the element, added which can be expressed by:

$$\sum_j \nu_j = 0 \qquad [2.14]$$

– if the component added is a reagent, its algebraic stoichiometric coefficient is negative and relation [2.13] yields $\Re' > 0$. The equilibrium shifts from left to right in the formulation of the balance equation.

– if the component added is a product of the reaction, its algebraic stoichiometric coefficient is positive and relation [2.13] yields $\Re' < 0$. The equilibrium shifts from right to left of the balance equation.

Hence, in the case of perfect solutions, if the transformation does not cause a variation in the number of moles, the addition of one of the components in the equilibrium shifts that equilibrium to consume the component added.

*Second case*: the transformation is accompanied by a decrease in the number of moles, so:

$$\sum_j \nu_j < 0 \qquad [2.15]$$

– if the component added is a product of the reaction, its algebraic stoichiometric coefficient is positive and relation [2.13] leads to $\Re' < 0$, so the system evolves so as to consume the component added;

– if the component added is a reagent, its algebraic stoichiometric coefficient is negative and relation [2.13] yields $\Re' > 0$ if:

$$\frac{|\nu_i|}{x_i} > \left|\sum_j \nu_j\right| \qquad [2.16]$$

or:

$$x_i < \frac{|\nu_i|}{\left|\sum_j \nu_j\right|} \qquad [2.17]$$

However, if:

$$|\nu_i| \geq \left|\sum_j \nu_j\right| \qquad [2.18]$$

the condition [2.17] is always satisfied because the molar fraction is always less than 1.

On the other hand, if:

$$|\nu_i| < \left|\sum_j \nu_j\right| \qquad [2.19]$$

The system evolves in the direction of consumption of the product added only if condition [2.14] is satisfied. On the other hand, if condition [2.14] is not fulfilled, the system reacts in the opposite direction, meaning that it produces the component added.

This case is encountered during the synthesis of ammonia by the reaction:

$$N_2 + 3H_2 = 2NH_3 \qquad [2R.1]$$

because in this case, we have:

$$\sum_j \nu_j = -2$$

If we add hydrogen, the equilibrium always shifts to the right. However, if we add nitrogen, the equilibrium shifts to the right only in the case where, in the initial equilibrium state, according to relation [2.14], we have a molar fraction of nitrogen less than 0.5.

*Third case*: if we have:

$$\sum_j \nu_j > 0 \qquad [2.20]$$

– if the component added is a reagent, its algebraic stoichiometric number is negative, and relation [2.14] leads to $\Re' > 0$ if:

$$\frac{\nu_i}{x_i} < \sum_k \nu_k \qquad [2.21]$$

This condition is always satisfied because we have:

$$\nu_i < 0,\ x_i > 0 \ \text{and} \ \sum_j \nu_j > 0$$

If the component added is a product of the reaction, its stoichiometric coefficient is positive, and we shall have $\Re' > 0$ if:

$$\frac{\nu_i}{x_i} > \sum_j \nu_j \qquad [2.22]$$

The system will evolve so as to produce the component added.

In the opposite case, the reaction will consume the component added.

Thus, we can definitively see that an equilibrium reacts in a complex way to the isobaric and isothermic addition of a component and that this behavior must be studied in each particular case. This study may become tricky in the case of non-perfect real solutions, for which the expressions of the derivatives of the chemical potentials in relation to the quantity of the component in question are complex.

### 2.1.5. *Influence of the addition of an inert component*

A component is said to be *inert* if it plays no part in the writing of the balance equation for the transformation:

– if that component is pure in its phase, its addition causes no alteration of the chemical potentials of the other components, which are null in that phase, and therefore no variation of the affinity. Consequently, that addition has no influence on the equilibrium, which is not modified;

– if the added component $A_i$ belongs to a polycomponent phase with other active components in the reaction, then relation [2.8] is simplified. If the stoichiometric coefficient of the component added is null, the condition of evolution becomes:

$$-\sum_j \nu_j \frac{\partial \mu_k}{\partial n_i} \Re' > 0 \qquad [2.23]$$

The sum is extended to the lone species intervening in the balance equation.

The result is still complex. If we restrict ourselves to the case of perfect solutions, then:

$$\frac{RT}{\sum_j n_j}\left(\sum_j \nu_j\right)\Re' > 0 \qquad [2.24]$$

– thus, if $\sum_j \nu_j$ is positive, the addition of $A_i$ yields $\Re' > 0$ and therefore the equilibrium evolves from left to right of the balance equation because of the addition of the inert component.

– if, on the other hand, $\sum_j \nu_j$ is negative, the addition of $A_i$ leads to $\Re' < 0$ and thus the equilibrium evolves from right to left of the balance equation because of the addition of the inert component.

## 2.2. Properties of all the equilibria in a system

We now consider a system in which various transformations take place simultaneously. Let $R$ represent the number of those transformations. Each transformation is represented by a balance equation $E_\rho$ which, for the $\rho^{th}$ reaction, is written as:

$$0 = \sum_k \nu_{k_\rho} \mathcal{A}_k \qquad [E_\rho]$$

Certain components may have a stoichiometric coefficient of 0 in reaction $\rho$.

The affinity of each transformation is of the form:

$$\mathcal{A}_\rho = -\sum_k \nu_{k_\rho} \mu_k \qquad [2.25]$$

We have seen (section 1.7) that the condition of overall equilibrium required each transformation to have a null affinity.

### 2.2.1. *Property of the set of balance equations of a system*

We consider the set $\mathcal{E}_R$ in the mathematical sense of the term of the $R$ balance equations of the transformations that take place in the system:

$$\mathcal{E}_R = \{E_1, E_2, ..., E_\rho, ..., E_R\} \quad [2.26]$$

We shall demonstrate a very important property relative to balance equations for the reactions in a system when those reactions are at equilibrium.

THEOREM 2.1.– all the balance equations of reactions at equilibrium occurring in a system constituting a vector space.

For the sake of comfort, we often speak of the vector space of equilibria.

To demonstrate our theorem, we shall show that on the set $\mathcal{E}_R$, we can define an internal composition law of addition and an external multiplication law on the set of real numbers $R$.

#### 2.2.1.1. *Internal composition law of addition*

Consider two balance equations $E_1$ and $E_2$ from that set $\mathcal{E}_R$. For the reaction $E_1$ we have:

$$0 = \sum_k \nu_{k_1} \mathrm{A}_k \quad [2R.2]$$

and for reaction $E_2$:

$$0 = \sum_k \nu_{k_2} \mathrm{A}_k \quad [2R.3]$$

Let us choose the sum of $E_1$ and $E_2$, which are components of balance equation $E_\Sigma$ defined by:

$$E_\Sigma = E_1 \oplus E_2 \quad [2.27]$$

Constructed with the sum of the terms of $E_1$ and $E_2$ and such that:

$$0 = \sum_k \nu_{k_1} \mathrm{A}_k + \sum_k \nu_{k_2} \mathrm{A}_k \quad [2R.4]$$

$E_\Sigma$ still represents a balance equation because, according to the definition of the affinity, it is evident that the affinity $\mathscr{A}$ of $E_\Sigma$ is the sum $\mathscr{A}_1 + \mathscr{A}_2$ of the affinities of $E_1$ and $E_1$. It stems from this that if the affinities $\mathscr{A}_1$ and $\mathscr{A}_2$ are null, then the affinity $\mathscr{A}$ is also null.

The set ($\mathcal{E}_R$, $\oplus$) is a commutative group because with three elements there is associativity:

$$E_\Sigma = (E_1 \oplus E_2) \oplus E_3 = E_1 \oplus (E_2 \oplus E_3) \qquad [2.28]$$

and we can define a neutral element $E_0$ in the addition as being the balance equation:

$$0 = 0 \qquad [2R.5]$$

and:

$$E_1 \oplus E_0 = E_0 \oplus E_1 = E_1 \qquad [2.29]$$

Each element has a symmetrical element. Regardless of $E_i$ of ($\mathcal{E}_R$, $\oplus$), there exists $E_i'$ of ($\mathcal{E}_R$, $\oplus$) such that:

$$E_i \oplus E_i' = E_i' \oplus E_i = E_0 \qquad [2.30]$$

The addition is commutative because:

$$E_1 \oplus E_2 = E_2 \oplus E_1 \qquad [2.31]$$

Hence, there is indeed an internal composition addition law.

#### 2.2.1.2. *External multiplication law on the set of real numbers*

This law is defined by an application of the set product ($R \otimes \mathcal{E}$) toward $\mathcal{E}$, which, for every couple has an element of ($R \otimes \mathcal{E}$) associated ($\lambda.E$), an element of $\mathcal{E}$. Indeed, by multiplying each term in a balance equation by $\lambda$, we have:

$$0 = \sum_k \nu_{k_p} \lambda.\mathrm{A}_k \qquad [2.32]$$

However, if $\mathscr{A} = 0$, then $\lambda.A = 0$, so $(\lambda.E)$ still represents a balance equation at equilibrium. This application is such that:

– regardless of the value of $E$ in $\mathfrak{S}_R$, where 1 represents the neutral element of the multiplication on $R$:

$$1.E = E \quad [2.33]$$

– irrespective of $\alpha$, $\beta$ of $R$ and regardless of the value of $E$ of $\mathfrak{S}$, we have the following if $\otimes$ is the symbol of multiplication on $R$:

$$\alpha.(\beta.E) = (\alpha \otimes \beta).E \quad [2.34]$$

and:

$$(\alpha + \beta).E = \alpha.E \oplus \beta.E \quad [2.35]$$

Regardless of the values of $\lambda$ for R and $E_1$ and $E_2$ of $\mathfrak{S}$, we have:

$$\lambda.(E_1 \oplus E_2) = \lambda.E_1 \oplus \lambda.E_2 \quad [2.36]$$

Hence, the $\mathfrak{S}_R$ has an external multiplication law on the set of real numbers.

Thus, the set $\mathfrak{S}_R$, which has an internal composition law of addition and an external multiplication law on the set of real numbers, is a vector space.

### 2.2.2. *Linear combinations of balance equations*

As the set $\mathfrak{S}_R$ constitutes a vector space, any linear combination with a balance equation such that:

$$E = \lambda_1 E_1 \oplus \lambda_2 E_2 \oplus ... \oplus \lambda_\rho E_\rho \oplus ... \oplus \lambda_R E_R \quad [2.37]$$

is also a balance equation that represents a resulting transformation.

The coefficients $\lambda_\rho$ are *multiplicative coefficients* of the balance equations $\rho$ in the combination in question.

The affinity of the resulting transformation will be given by the same linear combination of the affinities of the combined transformations.

$$\mathscr{A} = -\sum_{\rho}\sum_{k} \lambda_\rho \nu_{k(\rho)} \mu_k \qquad [2.38]$$

This expression enables us to calculate the affinity of the new, resulting transformation – an affinity which is null if the system is at thermodynamic equilibrium.

### 2.2.3. *Base of the vector space of the balance equations – Jouguet criteria*

Because the set of balance equations at equilibrium constitutes a vector space, it is sufficient, in order to obtain all of them, to find a base for that vector space. The number of balance equations which constitute that base will give us the dimension of the vector space – i.e. the number of independent reactions in the system.

Two problems are posed in practice.

*First problem*: we have a list of $R$ balance equations of a system at equilibrium. Are they all independent?

In order to answer this question, let us construct the matrix ($N \times R$) of the algebraic stoichiometric coefficients $\nu_{k_\rho}$ of each component $k$ in each reaction $\rho$.

$$\begin{Vmatrix} \nu_{1_1} & \nu_{2_1} & \cdots & \nu_{N_1} \\ \nu_{1_2} & \nu_{2_2} & \cdots & \nu_{N_2} \\ \cdots & \cdots & \cdots & \cdots \\ \nu_{1_R} & \nu_{2_R} & & \nu_{N_R} \end{Vmatrix}$$

In order for all the balance equations to be independent, no rows of the above matrix should be obtainable by a linear combination of other rows. This means that this matrix must contain at least one determinant of a non-null rank $R$. This condition is called the *Jouguet criterion* or *criterion of independence of the reactions*.

NOTE 2.1.– It is clear that the above criterion supposes that the number of transformations ($R$) is less than or equal to the number of components $N$ (number of rows of the matrix less than or equal to the number of columns) because, otherwise, it would no longer be possible to define a determinant of non-null rank $R$.

*Second problem*: finding a base of the vector space for the balance equations of a system.

This problem is important because the determination of a base will enable us to calculate the dimension of the vector space, which is, specifically, the number of transformations, which plays a part in Gibbs' phase law, for determining the number of independent components (see section 2.3). This number can also be used (see Chapter 3) to calculate the associated values and the equilibrium constant of the overall transformation.

There are a variety of practical methods to deal with our problem – each of them adapted to the nature of the system under study. We shall discuss three cases:

*Systems composed of neutral molecules* – for such a system, the most practical and effective method is to first establish the list of all the chemical species present in the system. Then, for each species, we write its chemical reaction of synthesis on the basis of its simple monatomic elements. By linear combination, we then eliminate the simple monatomic substances that are not really present in the system, and finally we eliminate the balance equations including simple substances that cannot be eliminated.

Next, in systems with multiple phases, we need to add the transformations of passage of the chemical species from one phase to another, for each chemical species present in a couple of phases.

We shall illustrate this method by using an example where we find a base for the vector space of the balance equations of a system containing methane (gas), carbon dioxide (gas), carbon monoxide (gas) and graphite (solid carbon).

Let us first write the synthesis reactions for the three components present in the system:

$$C + 4H = CH_4$$

$$C + 2O = CO_2$$

$$C + O = CO$$

Algebraically, by linear combination, we eliminate the monoxygen, which is absent from the system (whereas the element carbon is present in the form of graphite). In order to do so, we consider the last two equations, and by multiplying the last equation by 2, we obtain:

$$2CO = CO_2 + C$$

We find ourselves with two reactions:

$$C + 4H = CH_4$$

$$2CO = CO_2 + C$$

The first reaction involves the element hydrogen, which is neither present in the system nor algebraically eliminable, so this equation is eliminated. As no chemical species is present in more than one phase at a time, there is no phase transfer to be taken into account. Ultimately, the balance equation:

$$2CO = CO_2 + C$$

constitutes a base for our vector space, which is one-dimensional. There is only one independent equilibrium in our system.

NOTE 2.2.– dioxygen can be ignored in certain experimental conditions and in view of certain problems. If experience undermines this hypothesis, then we must take this into consideration, as well as its synthesis reaction. The system then becomes two-dimensional with two independent reactions.

*Ionic aqueous solutions* – in the case of ionic solutions, we first draw up a list of all the species (ions and neutral molecules) present in the solution. Then, we write Bronsted's weak-acid–weak-base equilibria and the redox reactions, the complexation reactions and finally the reactions of precipitation of solid phases.

By way of example, we shall examine the search for independent reactions in a system containing the following species:

– Gas: $NH_3$;

– Aqueous solution: $Ag^+$, $Cl^-$+, $H_3O^+$, $OH^-$, $NH4^+$, $[Ag_2(NH_3)_3]^{2+}$, $H_2O$, $NH_3$(aqu);

– Solid: AgCl.

Let us write the weak-acid–weak-base equilibria within the solution:

$$H_3O^+ + OH^- = 2H_2O$$

$$3H_2O + 2Ag^+ + 3NH_4^+ = Ag_2(NH_3)_3^{2+} + 3H_3O^+$$

$$NH_4^+ + H_2O = NH_3(aqu) + H_3O^+$$

We verify the independence of these three equilibrium states. Indeed, on the basis of the matrix of the algebraic stoichiometric coefficients, which is written as:

$$\begin{Vmatrix} -1 & -1 & 2 & 0 & 0 & 0 & 0 \\ 3 & 0 & -3 & -2 & -3 & 1 & 0 \\ 0 & -1 & 0 & 0 & -1 & 0 & 1 \end{Vmatrix}$$

we can deduce a non-null determinant of rank 3 (Jouguet criterion), such as:

$$\begin{vmatrix} -1 & -1 & 2 \\ 3 & 0 & -3 \\ 0 & -1 & 0 \end{vmatrix} = 4 \neq 0$$

In addition, we can verify that any other reaction written solely with components of the solution can be deduced from the above three equilibria. Such is the case, for instance, with the reaction:

$$3NH_3(aqu) + 2Ag^+ = Ag_2(NH_3)_3^{2+}$$

To the above three equilibria, we need to add the following two equilibria of phase transfer:

$$NH_3(aqu) = NH_3(gas)$$

$$Cl^- + Ag^+ = AgCl\,(solid)$$

Thus, the dimension of our vector space is 5. There can be no transformation within the gaseous phase and the solid phase, because these phases have only one component.

*Crystalline solids* – in the case of crystalline solids, we can no longer identify components and chemical species (atoms, ions or molecules), because the same species can occupy different sites in the crystalline lattice, which thus constitutes two different components. In addition, to take account of some of the properties of solids, we also need to take account of certain entities which appear to be irregularities in the arrangement of the species in space. These irregularities constitute point defects. We then define what is known as *structure elements*, and it is these which play the role of components. We are then led to the method used for ionic solutions.

## 2.3. Phase laws

Phase laws are used to determine the number of independent variables in a system at equilibrium.

There are two phase laws:

– Gibbs' phase rule, which pertains to the external intensive values and the composition variables. This law is applicable both to open and to closed systems;

– Duhem's law, which relates to all the intensive and extensive variables, and is applicable only to closed systems.

### 2.3.1. *Reminder of Gibbs' phase rule*

Gibb's phase rule states that the variance of a system with $p$ external intensive physical variables, containing $c$ independent components and $\phi$ phases, is given by:

$$v_G = c + p - \varphi \qquad [2.39]$$

The number of independent components is the difference between the number of components in the system and the number of independent transformations – i.e. the dimension of the vector space of the balance equations. This number must be decreased by 1 in the case of the presence of ions, due to the electrical neutrality condition, and decreased by 1 in the case of the species belonging to a solid phase if the condition of conservation of the crystallographic sites is imposed.

Remember that the variance is the number of free variables in the system, among the $p$ external intensive physical variables and the molar fractions. This number gives us the *maximum number of degrees of freedom* of the system; that maximum number of degrees of freedoms may be reduced by the imposition of specific constraints. Gibbs' phase rule applies equally to open systems and closed systems, and pertains solely to the intensive variables.

### 2.3.2. Duhem's phase rule in closed systems

Duhem's phase rule enables us to define a Duhem variance, which applies only to closed systems and pertains to the external intensive variables, the composition variables and the quantities of matter. The goal is always to determine the number of free variables.

#### 2.3.2.1. *Closure conditions*

As before, we suppose that the system contains $N$ components in $\phi$ phases and is home to $R$ independent transformations. The variation, over an arbitrary length of time, of the quantity $n_k$ of one of the components $k$ in a phase in the system will be given as a function of the reactional extents $\xi_\rho$ of the different transformations by:

$$\mathrm{d}n_k = \sum_{\rho=1}^{R} \nu_{k_\rho} \,\mathrm{d}\xi_\rho \qquad [2.40]$$

By integrating between the initial time, when $n_k = n_k^0$, and $\xi_\rho = 0$, we find:

$$n_k = n_k^0 + \sum_{\rho=1}^{R} \nu_{k_\rho} \xi_\rho \qquad [2.41]$$

The fact that the system is closed yields $N$ relations of the form [2.40] – one for each component. This set of relations constitutes the *closure conditions* for the system.

These equations may take a different form if $n^{(\alpha)}$ represents the total quantity of matter in a phase α, and $x_k^{(\alpha)}$ the molar fraction of the component $k$ in that phase. We would then have:

$$n_k = x_k^{(\alpha)} n^{(\alpha)} \qquad [2.42]$$

and relation [2.40], which is representative of the closure conditions, becomes:

$$x_k^{(\alpha)} n^{(\alpha)} = n_k^0 + \sum_{\rho=1}^{R} \nu_{k_\rho} \xi_\rho \qquad [2.43]$$

One of the two relations [2.40] or [2.42] can be used to impose closure of a system.

#### 2.3.2.2. *Duhem's theorem*

Thus, the state of a closed system will be fully determined if we know:

– the physico-chemical state of the phases by their $p$ external intensive values and their compositions, giving us $p + N$ variables;

– the total quantities of matter in each phase, which include $\phi$ variables.

Between all these variables, there are a certain number of relations which are:

– the $\phi$ relations which express the fact that, in each phase, the sum of the molar fractions is equal to 1;

– the conditions of equilibrium between the $R$ independent transformations;

– the $N$ closure relations of the form of equation [2.40]. Beware, though: if we know the initial quantities of matter, these last relations give us the $R$ variables of the reactional extents of the transformations at equilibrium, which need to be added to values remaining to be determined.

Thus, in total we have $p+N+\varphi+R$ variables and $N+\varphi+R$ relations, so the degree of indeterminacy, if we know the initial quantity of each component, is:

$$v_D = p+N+\varphi+R-(N+\varphi+R)=p \qquad [2.44]$$

This brings us to Duhem's variance theorem:

THEOREM 2.2.– the equilibrium states of a closed system, wherein we know the initial quantities of matter, are completely determined if we set $p$ variables.

### 2.3.3. *Comparison between the Gibbs variance and the Duhem variance*

Consider a closed system. Gibbs' theorem enables us to determine the Gibbs variance $v_G$ pertaining solely to the intensive physico-chemical variables (external and compositional). Duhem's theorem gives a variance $v_D = p$ pertaining to both the intensive and extensive values of a closed system. Three cases may arise:

– if $v_G < v_D$ the Gibbs variance gives the number of free intensive physico-chemical variables, and the difference $v_D - v_G$, if the system is closed, gives the number of free extensive variables;

– if $v_G = v_D$, there are only intensive variables (external and compositional) that are free. The number thereof is determined by $v_G$;

– if $v_G > v_D$, the conditions of the closed system are automatically fulfilled simply by the application of Gibbs' phase rule.

In the case of systems composed solely of condensed phases, except at very high pressures, the pressure is no longer an equilibrium variable, which means that the value of $p$ decreases by 1.

## 2.4. Indifferent states

We shall now look at a particular class of states of a system: its *indifferent states*.

### 2.4.1. *Definition*

A state is said to be indifferent if it is possible to subject the closed system in that state to a virtual azeotropic displacement.

### 2.4.2. *Condition of indifference of a state*

As the initial state, take a system with $N$ components, home to $R$ independent transformations, characterized by $p$ external intensive variables ($P$, $T$, etc.), the molar fractions of the different components $A_k$ in each phase $\alpha$ ($x_k^{(\alpha)}$) and by the quantity of matter in the different phases $n^{(\alpha)}$.

Let us perform an azeotropic displacement (see section 1.9) of that state to bring it to the final state characterized by the same variables assigned the "prime" index. During the course of that transformation, as the system remains closed, it obeys the closure condition [2.45], which enables us to write:

$$x'^{(\alpha)}_k n'^{(\alpha)} - x_k^{(\alpha)} n^{(\alpha)} = \sum_{\rho=1}^{R} \nu_{k_\rho} \xi_\rho \qquad [2.45]$$

As the transformation is azeotropic, the application of relation [1.90] gives us:

$$x_k^{(\alpha)} = x'^{(\alpha)}_k \qquad [2.46]$$

Then, expression [2.45] becomes:

$$x_k^{(\alpha)} \left( n'^{(\alpha)} - n^{(\alpha)} \right) = \sum_{\rho=1}^{R} \nu_{k_\rho} \xi_\rho \qquad [2.47]$$

By setting the following for the purposes of simplicity:

$$\Delta_r n^{(\alpha)} = n'^{(\alpha)} - n^{(\alpha)} \qquad [2.48]$$

we obtain a system of $N$ equations in the form:

$$x_k^{(\alpha)} \Delta_r n^{(\alpha)} = \sum_{\rho=1}^{R} \nu_{k_\rho} \xi_\rho \qquad [2.49]$$

The unknowns in this system of equations are $\Delta_r n^{(\alpha)}$, $\Delta_r n^{(\beta)}$, …, $\Delta_r n^{(\varphi)}$, $\xi_1$, $\xi_2$, …, $\xi_R$, which is to say that there are $\phi + R$ unknowns. The original state will therefore be indifferent if the system of equations can have a solution such that the terms $\Delta_r n^{(\alpha)}$ are not all null. However, the Gibbs variance of that system can be written as:

$$v_G = N - (\varphi + R) + p \qquad [2.50]$$

The indifference conditions [2.46] will therefore depend on the difference between the number of equations $N$ and the number of unknowns $\phi + R$; and depending on the values of the variance, two cases need to be considered:

1) If $v_G < p$, then $N < (\varphi + R)$ and the system of equations [2.49] will always have fewer equations than unknowns. Thus it will have solutions different to zero and we can arbitrarily choose a certain number of nonzero values $\Delta_r n^{(\alpha)}$ – that number is $(p - v_G)$. Hence, systems whose Gibbs variance is less than the number of external intensive parameters $p$ are always indifferent.

2) If $v_G \geq p$, then $N \geq (\varphi + R)$. In order for the system of equations [2.49] to have an acceptable solution (at least one nonzero value of $\Delta_r n^{(\alpha)}$) it is necessary and sufficient for the determinants of order $\varphi + R$, formed on the basis of the unknowns, to have the value of 0. The corresponding matrix is written as:

$$\begin{Vmatrix} x_{1,1} & x_{1,2} & \cdots & x_{1,\varphi} & v_{1,1} & v_{1,2} & \cdots & v_{1,R} \\ x_{2,1} & x_{2,2} & \cdots & x_{2,\varphi} & v_{2,1} & v_{2,21} & \cdots & v_{2,R} \\ \cdots & \cdots & \cdots & \cdots & \cdots & \cdots & \cdots & \cdots \\ x_{N,1} & x_{N,1} & \cdots & x_{N,\varphi} & v_{N,1} & v_{N,2} & \cdots & v_{N,R} \end{Vmatrix} \qquad [2.51]$$

As there are several determinants which must be null, the number of indifference conditions will depend on the dimensions of matrix [2.51].

If $N = (\varphi + R)$, there will be only one determinant, and therefore only one indifference condition.

### 2.4.3. *Set of indifferent points of equilibrium*

The indifference conditions of a system are unrelated to the states of equilibrium of that system. This means that certain indifferent states could possibly be equilibrium states, but there are also indifferent states which are not equilibrium states.

We have seen that in intensive variables, the equilibrium states constitute a set of dimension $v_G$ (Gibbs variance); the elements in that set also belong to the set of indifferent states, and therefore, satisfying $v_G + p - 1$ additional conditions will constitute a set of dimension: $v_G + p - 1 - v_G = p - 1$. Therefore, we can state Saurel's theorem:

THEOREM 2.3.– The subset of the indifferent equilibrium states of a system is represented by a set of dimension $p - 1$, independent of the variance of the system.

For example, by bringing together the conditions of equilibrium and those of indifference, we can show that in a monovariant system with two external intensive variables ($P$ and $T$), all the equilibrium states are indifferent states.

### 2.4.4. *Gibbs–Konovalov theorem*

We shall now look at a theorem developed by Gibbs and Konovalov. For the moment, we shall simply accept this theorem as true. We shall demonstrate it later when we examine equilibria between phases.[1]

THEOREM 2.4.– In a system with two external intensive variables – pressure and temperature – at equilibrium, for any shift at constant temperature (or respectively at constant pressure), the pressure (or respectively the temperature) reaches an extremum for an indifferent state, and vice versa. If amongst the values of the pressure (or respectively the temperature) which keep the system at equilibrium at constant temperature (or respectively constant pressure), there is an extremum value, then the state corresponding to that value is indifferent.

1 See Volume 5 of this Set of books: *Phase Transformations*.

NOTE 2.3.– Obviously, this theorem only makes sense if the Gibbs variance is at least equal to 2.

## 2.5. Thermodynamically-equivalent systems

Consider one closed chemical system A. It is composed, to begin with, of a mixture of a certain number of components in known quantities. This system evolves spontaneously at constant pressure and temperature. In general, with the exception of oscillating systems, this system approaches a state of equilibrium. In order for this to happen, a number of reactions occur which, at any given moment, are each characterized by their fractional extent $\xi_i$. At equilibrium, each reaction reaches a limit extent or equilibrium extent $\xi_i^{(e)}$. Now we construct another system, B, mixing the same components as before, with the quantities identical to those obtained when the reactions in system A have each attained a given fractional extent $\xi_i$. This system B will, under the same conditions of temperature and pressure, obviously attain the same state of equilibrium as the previous system. We say that *the two systems A and B are thermodynamically equivalent.* We can even state that the second system, B, is closer to equilibrium than system A.

The concept of thermodynamically-equivalent systems is used, in particular, when studying sets of several reactions, using the predominant-reaction method. The method is very frequently used for calculating the state of equilibrium of ionic reactions in an aqueous solution – e.g. calculating the pH of a solution.

## 2.6. Stability of equilibria

The concept of stability of a state of equilibrium is a relative notion. That stability is understood in relation to a specific transformation. For example, if we take oxygenated water at standard temperature and pressure (STP), we know that this substance is unstable in terms of water but is stable in terms of the hydrogen–oxygen mixture.

### 2.6.1. *De Donder's general stability condition*

Consider a system $S$ defined by the chosen values of the intensive variables and by an extent $\xi$ of the transformation envisaged ($\xi$ can be null). The values of the different physical intensive variables are kept constant. If, at a given time, that system is at equilibrium, it is because its rate $\Re_S$ is null. At the same time, imagine another system $S$', identical to $S$ except for its extent, which is $\xi' = \xi + \delta\xi$. System $S$' is said to be disturbed in relation to system $S$. It is generally not at equilibrium, and thus its rate $\Re_{S'}$ is different to zero. $\delta\xi$ is called the disturbance.

De Donder says that system $S$ is stable in relation to the disturbance $\delta\xi$ if the rate of system $S$' has the sign which tends to bring $S$' back to state $S$.

Hence if $\delta\xi > 0$, then $\Re_{S'}$ must be negative in order for system $S$ to be stable. Conversely, if $\delta\xi < 0$, then $\Re_{S'}$ must be positive. This gives us the stability condition of $S$ according to de Donder:

$$\Re_{S'}.\delta\xi < 0 \tag{2.52}$$

### 2.6.2. *Stability of a system with unilateral variations*

System $S$ will be said to be of *unilateral variations* if the disturbance cannot take only a single positive or negative sign. Thus, we can take the origin $\xi = 0$ for the extent.

Let us choose, for example, $\delta\xi > 0$ (we could also have made the opposite choice and the conclusions drawn would have been the same). In view of the inequality [2.52], system $S$ will be stable if:

$$\Re_{S'} < 0 \tag{2.53}$$

Yet system $S$' obeys de Donder's inequality [1.50], so here:

$$\mathscr{A}_{S'}\Re_{S'} > 0 \tag{2.54}$$

The stability condition [2.53] becomes:

$$\mathrm{A}_{S'} < 0 \qquad [2.55]$$

However, we can write:

$$\mathcal{A}_{S'} = \mathcal{A}_S + \left(\frac{\partial \mathcal{A}_S}{\partial \xi}\right)_{\xi=0} \mathrm{d}\xi \qquad [2.56]$$

Two cases may therefore arise.

– if system $S$ is at thermodynamic equilibrium, then $\mathcal{A}_S = 0$ and therefore:

$$\mathcal{A}_{S'} = \left(\frac{\partial \mathcal{A}_S}{\partial \xi}\right)_{\xi=0} \mathrm{d}\xi \qquad [2.57]$$

Hence, we have the following conditions of stability of $S$:

$$\mathcal{A}_S = 0 \qquad [2.58a]$$

and:

$$\left(\frac{\partial \mathcal{A}_S}{\partial \xi}\right)_{\xi=0} < 0 \qquad [2.58b]$$

– if the system is at false equilibrium, then $\Re_S = 0$ $\mathcal{A}_S \neq 0$; if we ignore the second term in expression [2.56], then the condition of stability of $S$ becomes:

$$\Re_S = 0 \qquad [2.59a]$$

and:

$$\mathcal{A}_S < 0 \qquad [2.59b]$$

NOTE 2.4.– These conditions can, of course, very easily be expressed in terms of chemical potentials by using expressions [1.23] and [1.42].

NOTE 2.5.– These conditions can also be expressed in terms of characteristic functions. Then by using expression [1.20], if the system is at thermodynamic equilibrium, condition [2.58] is written:

$$\frac{\partial \Gamma}{\partial \xi} = 0 \qquad [2.60a]$$

and:

$$\frac{\partial^2 \Gamma}{\partial \xi^2} < 0 \qquad [2.60b]$$

The concavity of the curve $\Gamma(\xi)$ is turned toward positive values of $\Gamma$ and exhibits a minimum at the origin. Therefore, for chemical systems, as function $\Gamma$ is the Gibbs energy function $G$, the curve showing this Gibbs energy as a function of the extent of the reaction has a minimum at the origin.

If the system is at false equilibrium, then condition [2.59] becomes:

$$\frac{\partial \Gamma}{\partial \xi} \neq 0 \qquad [2.61a]$$

and:

$$\frac{\partial^2 \Gamma}{\partial \xi^2} > 0 \qquad [2.61b]$$

The curve $\Gamma(\xi)$ does not have a minimum at the origin.

### 2.6.3. *Stability of a system with bilateral variations*

The system will be subject to *bilateral variations* if the disturbance $\delta\xi$ can have any sign. Let us show that such a system, if it is stable, is necessarily thermodynamically stable ($\mathscr{A}_S = 0$).

Indeed, suppose that it is otherwise, and that $\mathscr{A}_S > 0$, for example. De Donder's inequality for the disturbed system, if we discount the variation of the affinity, gives us:

$$\mathscr{A}_S \Re_{S'} > 0 \qquad [2.62]$$

Hence:

$$\Re_{S'} > 0 \quad [2.63]$$

If we choose a positive value for $\delta\xi$, then the product $\Re_{S'}.\delta\xi$ is positive, which runs counter to the stability condition [2.52]. We would find the same conclusion by choosing $\mathscr{A}_S < 0$ , which would give us $\mathscr{A}_S = 0$ .

In this case, the affinity of the disturbed system will be:

$$\mathscr{A}_{S'} = \left(\frac{\partial \mathscr{A}_S}{\partial \xi}\right)\delta\xi \quad [2.64]$$

and according to de Donder, we should have:

$$\frac{\partial \mathscr{A}_S}{\partial \xi}\Re_{S'}.\delta\xi > 0 \quad [2.65]$$

Therefore, the stability condition [2.50] is imposed:

$$\frac{\partial \mathscr{A}_S}{\partial \xi} < 0 \quad [2.66]$$

Hence, for a system with bilateral variations to be stable, it is necessary and sufficient that the following two conditions be fulfilled:

$$\mathscr{A}_S = 0 \quad [2.67a]$$

and:

$$\frac{\partial \mathscr{A}_S}{\partial \xi} < 0 \quad [2.67b]$$

Using relation [1.20], in order for a system with bilateral variations to be stable, it is necessary and sufficient for the characteristic function to obey the following two conditions:

$$\frac{\partial \Gamma}{\partial \xi} = 0 \quad [2.68a]$$

and:

$$\frac{\partial^2 \Gamma}{\partial \xi^2} > 0 \qquad [2.68b]$$

Thus, the curve $\Gamma(\xi)$ exhibits a minimum.

NOTE 2.6.– It may be that the first derivative (and even some of the subsequent derivatives) of the affinity in relation to the extent will have a value of 0, in which case we must continue the expansion [2.64] until we find the first nonzero derivative. We then show that the system cannot be stable if the first nonzero derivative is of an even order. The stability condition becomes:

$$\frac{\partial^{2n+1} \mathscr{A}_S}{\partial \xi^{2n+1}} < 0 \qquad [2.69a]$$

or:

$$\frac{\partial^{2n+1} \Gamma}{\partial \xi^{2n+1}} > 0 \qquad [2.69b]$$

with all the derivatives of order less than $2n$+1 having a value of 0.

### 2.6.4. *Conditions of bilateral stability expressed in terms of chemical potentials*

If we express the stability condition [2.67] using expressions [1.23] and [1.42], we obtain:

$$\frac{1}{2}\sum_i \sum_k \frac{\partial \mu_i}{\partial n_k}\left(\frac{v_i}{n_i} - \frac{v_k}{n_k}\right)^2 n_i n_k < 0 \qquad [2.70]$$

This inequality thus leads to the condition:

$$\frac{\partial \mu_i}{\partial n_k} < 0 \text{ for } i \neq k \qquad [2.71]$$

Thus, the condition of bilateral stability can be expressed very simply in terms of chemical potentials.

# 3

# Molecular Chemical Equilibria

In this chapter, we shall apply the general principles of transformations at equilibrium, which we examined in Chapter 1, to the case of chemical reactions. Hence, we define the equilibrium constant, and its different variants, the value of which we shall see how to determine in Chapter 4.

Numerous graphical representations are used to represent equilibrium states. The main ones which are used for molecular reactions have been included in this chapter.

## 3.1. Law of mass action – equilibrium constants

In this section, we introduce the equilibrium constant, its different forms and its properties.

### 3.1.1. *Expression of the law of mass action*

The law of mass action is obtained from expression [1.29], and by writing that the affinity is null at thermodynamic equilibrium.

Immediately we can deduce that at equilibrium, the reaction quotient takes the particular value $Q_r^{(\text{equ})}$, and therefore the equilibrium condition is:

$$\mathrm{R}T \ln Q_r^{(\text{equ})} = \mathscr{A}_r^0 \qquad [3.1]$$

This particular value of the reaction quotient at equilibrium is known as the *equilibrium constant*. Thus, it can be expressed as a function of the affinity in the reference state or, by virtue of relation [1.25], of the Gibbs energy associated with the reaction in the reference state:

$$K_r = Q_r^{(\text{equ})} = \exp\frac{\mathscr{A}_r^0}{\text{R}T} = \exp-\frac{\Delta_r G^0}{\text{R}T} \qquad [3.2]$$

The reaction conditions and equilibrium conditions of our system will thus be easily expressed in terms of the value of the reaction quotient. Obviously, we shall have:

– if $\mathscr{A}_r > 0$, or $\Delta_r G < 0$, then $Q_r^{(\text{equ})} < K$ and the reaction occurs spontaneously from left to right;

– if $\mathscr{A}_r > 0$ or $\Delta_r G > 0$, then $Q_r^{(\text{equ})} > K$ and the reaction is impossible, so the opposite reaction occurs spontaneously;

– if $\mathscr{A}_r = 0$ or $\Delta_r G = 0$ then $Q_r^{(\text{equ})} = K$, and the reaction is at thermodynamic equilibrium.

NOTE 3.1.– The equilibrium constant, which depends only on the Gibbs energies in the reference state, is independent of the state of perfection of the phases. This means that this constant does not depend on the nature of the mixture of the reagent species, but only on the nature of the products. For example, whether we are studying the carbon-oxygen-carbon monoxide balance using pure carbon or carbon in solution in iron (steel), the equilibrium constant is the same in both cases, because we can choose carbon in a pure phase as a reference state.

By comparing the first equality in relation [3.2] and expression [1.28], the law of mass action becomes:

$$K_r = \prod_i a_i^{\nu_i} \qquad [3.3]$$

Obviously, the values of the activities are those which reign at thermodynamic equilibrium, and no longer values in any given state, as was the case with relation [1.28].

NOTE 3.2.– the activities and values of the reference properties depend on the convention chosen in the description of the solution. It follows that the equilibrium constant will depend on the convention chosen for describing the solution. Thus, *a priori*, we shall have three possible values for an equilibrium constant, which we shall write as: $K^{(\mathrm{I})}$, $K^{(\mathrm{II})}$ and $K^{(\mathrm{III})}$.

If we adopt the pure-substance reference (I), and if the solution is perfect, the activity coefficient is equal to 1 and the law of mass action takes the form of relation [3.3a], replacing the activities in that formula with the molar fractions, as follows:

$$K_r^{(\mathrm{I})} = \prod_i x_i^{\nu_i} \qquad [3.3a]$$

In that case, the Gibbs energy of the reference state is the standard Gibbs energy of the reaction at the temperature at hand, i.e. $\Delta_r g^0$. Remember that the value of a function for a substance in the standard state at temperature $T$ is its value for a pure substance, in its normal state of condensation at a pressure of 1 bar and at temperature $T$.

### 3.1.2. *Different forms of the law of mass action*

Depending on the values chosen to write the compositions of the multi-component phases, the expression of the reaction quotient may take a variety of forms, and therefore there are also different possible forms for the law of mass action.

If we have a homogeneous system in the gaseous phase, it is preferable to quantify the compositions by the partial pressures, and the law of mass action will be expressed in terms of fugacity:

$$\prod_k \left(\frac{f_k}{P^0}\right)^{\nu_k} = K^{(f)} \qquad [3.4]$$

The constant $K^{(f)}$ is called the equilibrium constant relative to the fugacities.

There is an extremely simple specific case of the calculation of the fugacity of the gas *i*: when we consider that the mixture of gases forms a perfect solution of imperfect gases, because then we have $\gamma_i^{(\mathrm{I})} = 1$, the fugacity is written as:

$$f_i = f_i^0 y_i = \varphi_i^0 y_i P \qquad [3.5]$$

This is what is known as Lewis' statement. $\varphi_i^0$ is the fugacity coefficient of the pure gas *i* at the same total pressure as in the mixture. In this particular case, albeit one which is encountered frequently, by substituting back into equation [3.4] we obtain:

$$\prod_i \left( \frac{\varphi_i^o P_i}{P^0} \right)^{\nu_i} = K^{(f)} \qquad [3.6]$$

In the ideal case of perfect solutions of perfect gases, the fugacity coefficient for each of the gases is equal to 1, and the fugacity is identical to the partial pressure. The law of mass action then becomes:

$$K^{(P)} = \prod_i \left( \frac{P_i}{P^0} \right)^{\nu_i} \qquad [3.7]$$

$K^{(P)}$ is the *equilibrium constant relative to the partial pressures*. The reference pressure $P^0$ is generally taken to be 1 bar.

In order to link the equilibrium constant relative to the partial pressures to the equilibrium constant *K* defined by relation [3.3], which pertains to the molar fractions, because we are dealing with perfect solutions we shall base our arguments on the law of mass action [3.7] relative to perfect gases, given that the equilibrium constants are defined on the basis of the standard conditions, which depend solely on the species in the pure state. We express the partial pressure as a function of the molar fraction, by:

$$P_i = P x_i \qquad [3.8]$$

The application of relation [3.8] enables us to write:

$$\prod_i x_i^{\nu_i} \left( \frac{P}{P^0} \right)^{\sum_i \nu_i} = K^{(P)} \qquad [3.9]$$

Thus, we find the sought relation:

$$K^{(P)} = K^{(\mathrm{I})}\left(\frac{P}{P^0}\right)^{\sum_i \nu_i} \qquad [3.10]$$

Using expressions [3.2] and [3.10], the equilibrium constant relative to the partial pressures takes the form:

$$K^{(P)} = \left(\frac{P}{P^0}\right)^{-\sum_i \nu_i} \exp\frac{\mathscr{A}_r^o}{\mathrm{R}T} \qquad [3.11]$$

NOTE 3.3.– If the system is studied at a total pressure equal to the reference pressure $P^0$ (normally 1 bar), the two constants $K^{(P)}$ and $K$ are identical.

If we now choose to quantify the composition of the gaseous phase in terms of the concentrations, we write the following for a perfect gas:

$$P_k = C_k \mathrm{R}T \qquad [3.12]$$

By substituting this back into relation [3.7], we obtain:

$$\prod_i \left(\frac{C_i}{C_i^0}\right)^{\nu_k} (\mathrm{R}T)^{\sum_i \nu_i} = K^{(P)} = (\mathrm{R}T)^{\sum_i \nu_i} K^{(c)} \qquad [3.13]$$

From relation [3.13], we deduce that the equilibrium constant relative to the concentrations is linked to the equilibrium constant relative to the molar fractions by:

$$K^{(c)} = \prod_i \left(\frac{C_i}{C_i^0}\right)^{\nu_k} = (\mathrm{R}T)^{-\sum_i \nu_i}\left(\frac{P}{P^0}\right)^{\sum_i \nu_i} K^{(\mathrm{I})} \qquad [3.14]$$

Very often, the reference concentration $C_i^0$ is taken as 1 mole per liter for each component, and if the concentrations are expressed in moles per liter, the law of mass action takes the form:

$$K^{(c)} = \prod_i (C_i)^{\nu_i} \qquad [3.15]$$

If, now, the equilibrium is obtained in a condensed phase, we need to consider the particular – and very commonplace – case of highly-dilute solutions, where the solvent is one of the reagents. This situation is frequently encountered with dilute aqueous solutions. In such cases, the solvent has a chemical potential which practically depends only on the temperature and is given by: $\mu_0^* = g_0^0$; in addition, its activity is practically one ($a_0 \cong 1$), so that the law of mass action is written by using relation [3.3], in which the term relating to the solvent in the right-hand side no longer plays a part. Note, however, that the term relating to the solvent always comes into play on the left-hand side of the equation, i.e. in the definition of the equilibrium constant according to the last equality in relation [3.2].

Consider a dilute solution of solutes $A_i$ in a solvent $A_0$. Let $v_0^0$ denote the molar volume molar of the pure solvent, and $x_0$ and $x_i$ the respective molar fractions of the solvent and solute $A_i$.

If we look again at relation [3.2], applied to the reference state (III) – molar solution – we have:

$$K^{(\mathrm{III})} = \exp - \frac{\Delta_r \mu^{0(\mathrm{III})}}{RT} \qquad [3.16]$$

However, in the reference solution, in view of relation [A1.16] (see Appendix 1), the chemical potentials of reference state (III) and the infinitely-dilute reference state (II) are equal, so the equilibrium constant is the same in both cases:

$$K^{(\mathrm{III})} = K^{(\mathrm{II})} \qquad [3.17]$$

If we now combine relations [A1.5] and [3.17], we obtain the following (only the solutes are involved):

$$\prod_s \left(\gamma_s^{(\mathrm{III})} C_s\right)^{\nu_s} = \prod_s \left(\gamma_s^{(\mathrm{II})} x_s\right)^{\nu_s} = K^{(\mathrm{II})} \qquad [3.18]$$

If the reference concentration is 1 mole/L, then in a sufficiently-dilute solution, in light of relation [A1.18], we obtain an equilibrium constant relative to the concentrations which depends on the molar volume of the solvent:

$$\prod_s C_s^{\nu_s} = K^{(\mathrm{II})}\left(v_0^0\right)^{\sum_s \nu_s} \qquad [3.19a]$$

NOTE 3.4.– If the reaction takes place with the number of moles being constant (i.e. $\sum_s \nu_s = 0$), we can regard the activities and concentrations as one and the same thing, and write:

$$\prod_s C_s^{\nu_s} = \prod_s a_s^{\nu_s} = K^{(\mathrm{II})} \qquad [3.19b]$$

Let us now examine the application of the law of mass action if we choose to use reference (II) in solution, and compare it to the equilibrium constant obtained in reference (I). By using relation [A1.13], which introduces the Henry's constant for each component, and substituting the result back into relation [3.3], we obtain:

$$K^{(\mathrm{I})} = \prod_s a_s^{\nu_s(\mathrm{I})} = \prod_s a_s^{\nu_s(\mathrm{II})} K_{sH}^{\nu_s} \qquad [3.20]$$

This gives us the relation between the equilibrium constant relative to the pure-substance reference (I) and that relative to the infinitely-dilute solution reference (II):

$$K^{(\mathrm{II})} = \frac{K^{(\mathrm{I})}}{\prod_s K_{sH}^{\nu_s}} \qquad [3.21]$$

NOTE 3.5.– for a perfect solution, given that the Henry's constant of the different components are all equal to one, the equilibrium constants $K^{(\mathrm{II})}$ and $K^{(\mathrm{I})}$ are identical. As, according to relation [3.17], the equilibrium constants in reference (II) and reference (III) are identical, it follows that, for a perfect solution, all the equilibrium constants are identical.

### 3.1.3. *Use of solution models and application of the law of mass action*

The use of the law of mass action to predict equilibrium states for reactions where all the components are in the same solution poses two problems:

– The first is the knowledge of the activity coefficients. They cannot be determined experimentally, because we are unable to take measurements on the solution, other than in conditions of equilibrium, because of the reaction itself which is taking place. Hence, in general, we need to use physical models or mathematical models obtained by extrapolation from measurements taken in other conditions.

– The second problem stems from the fact that the calculation of the activity coefficients, using models, requires us to know the concentrations at equilibrium – concentrations which we can only obtain by dosages or by application of the law of mass action if we know those activity coefficients.

In the latter case, we proceed iteratively. Beginning with a perfect system or an ideal dilute system with activity coefficients of one, we calculate a first series of values of the concentrations, starting with a known initial state. Then, using that model, we calculate the activity coefficients at the concentrations thus obtained. Again applying the law of mass action gives us a second series of values for the concentrations; on the basis of those values we calculate a second series of activity coefficients, and so on until we obtain a correct convergence.

Numerous reactions take place in polyphase media and involve components belonging to different phases. Thus, we may simultaneously encounter gases, species in solution and/or substances in pure phases (essentially solids), which are involved in the same equilibrium. Thus, the expression of the law of mass action will involve the fugacities (or partial pressures) for the gases, the activities (or the molar fractions) for the species in solution. There will be no term relating to a component in a pure phase. Of course, the equilibrium constant will contain all the terms relating to all the components in the reaction in the definition of the form [3.2] as a function of the Gibbs energy of the chosen reference state.

### 3.1.4. *Systems composed of a set of equilibria*

We now consider a system comprising a set of equilibria. As the set of balance equations of these reactions constitutes a vector space $\mathfrak{E}$ (see section 2.2.1), we can deduce that any linear combination of a certain number of balance equations $E_1$, $E_2$, … $E_m$, such that:

$$E_r = \lambda_1.E_1 \oplus \lambda_2.E_2 \oplus ... \oplus \lambda_m.E_m \qquad [3.22]$$

is also a balance equation for the system.

For the expressions of the different affinities of the corresponding reactions – e.g. for a reaction $\rho$ in which the algebraic stoichiometric number of the component $A_i$ is $\nu_{i_\rho}$ – we have:

$$\mathscr{A}_\rho = -\sum_k \nu_{i_\rho} \mu_i \qquad [3.23]$$

Thus, the affinity of the reaction corresponding to the balance equation $E_r$ will be:

$$\mathscr{A} = \sum_\rho \lambda_\rho \mathscr{A}_\rho = -\sum_\rho \sum_i \lambda_\rho \nu_{i_\rho} \mu_i \qquad [3.24a]$$

A similar relation can be established for the reference affinity of reaction $E_r$:

$$\mathscr{A}^0 = \sum_\rho \lambda_\rho \mathscr{A}_\rho^0 = -\sum_\rho \sum_i \lambda_\rho \nu_{i_\rho} \mu_i^0 \qquad [3.24b]$$

The equilibrium constants are linked to the reference affinities by relations of the form:

$$\mathrm{R}T \ln K_\rho = \mathscr{A}_\rho^0 \qquad [3.25]$$

Relation [3.24] enables us to write the following for the equilibrium constant of the reaction $E_r$:

$$\mathrm{R}T \ln K_r = \mathrm{R}T \sum_\rho \lambda_\rho \ln K_\rho \qquad [3.26]$$

This gives us the relation between the equilibrium constant $K_r$ and those of the combined reactions $K_\rho$:

$$K_r = \prod_\rho (K_\rho)^{\lambda_\rho} \quad [3.27]$$

This expression enables us to calculate the constant of any given equilibrium, if we know the equilibrium constants of the combined reactions, in a chosen convention to describe the solutions.

### 3.1.5. *Unit of the equilibrium constants*

Irrespective of the reaction under study and the expression of the law of mass action used, the corresponding equilibrium constant has no unit, because it is always defined by an exponential. This means, in particular, that the law of mass action expressed in partial pressures must contain only the pressure ratios. Similarly, in terms of concentrations, as we have seen, they play a part in the ratios of concentrations. It is true that the denominator of those fractions is often 1 (normal pressure of 1 bar and reference concentration 1 mole/l).

However, it must be remembered that the equilibrium constant depends on the convention chosen to define the solution.

### 3.1.6. *Variations of the equilibrium constants with temperature*

By applying relation [A1.22], by deriving relation [3.2] in relation to temperature:

$$\frac{\partial \ln K}{\mathrm{d}T} = \frac{\mathrm{d}\left(\frac{\mathscr{A}_r^0}{\mathrm{R}T}\right)}{\mathrm{d}T} = \frac{\Delta_r H^0}{\mathrm{R}T^2} \quad [3.28]$$

By deriving relation [3.10], we obtain:

$$\frac{\partial \ln K^{(P)}}{\mathrm{d}T} = \frac{\partial \ln K^{(\mathrm{I})}}{\mathrm{d}T} = \frac{\Delta_r h^0}{\mathrm{R}T^2} \quad [3.29]$$

$\Delta_r h^0$ pertains to the enthalpies of the pure substances, because the reference state is that which chooses the pure substances.

Similarly, by deriving relation [3.14], we obtain the following, in the case of gases:

$$\frac{\partial \ln K^{(c)}}{\partial T} = \frac{\partial \ln K^{(\mathrm{I})}}{\partial T} - \frac{\sum_i \nu_i}{T} \quad [3.30]$$

Hence, considering relation [3.28]:

$$\frac{\partial \ln K^{(c)}}{\partial T} = \frac{\Delta_r h^0 - \mathrm{R}T\sum_i \nu_i}{\mathrm{R}T^2} \quad [3.31]$$

For perfect gases, the enthalpy is independent of the pressure and the internal energy does not depend on the volume:

$$\Delta_r h^0 = \Delta_r u^0 + \Delta_r \left(P v_0^0\right) \quad [3.32]$$

For reactions involving perfect gases:

$$\Delta_r \left(P v_0^0\right) = -\mathrm{R}T \sum_i \nu_i \quad [3.33]$$

By combining relations [3.32] and [3.33], we obtain:

$$\Delta_r h^0 = \Delta_r u^0 + \mathrm{R}T \sum_i \nu_i \quad [3.34]$$

and thus, by substituting back into relation [3.31], we obtain the following for perfect gases:

$$\frac{\partial \ln K^{(c)}}{\partial T} = \frac{\Delta_r u^0}{\mathrm{R}T^2} \quad [3.35]$$

$\Delta_r u^0$ is the Helmholtz energy associated with the reaction $r$ in the standard state. This result constitutes the van't Hoff isochore.

NOTE 3.6.– it must be carefully noted that relation [3.35] applies only to mixtures of gases, and is not demonstrated for reactions in a condensed solution.

### 3.1.7. *Influence of the choice of reference pressure on the equilibrium constant*

The equilibrium constant of partial pressures does not depend on the pressure in the experiment, but instead depends on the reference pressure. However, we may be led to change the reference pressure, which leads to an alteration of the value of $K_r^{(P)}$. This occurs, in particular, when we use tables of old thermodynamic data (before 1980), for which the reference pressure was that of the standard state chosen theoretically at 1 atmosphere (i.e. $1.129 \times 10^5$ Pa) instead of 1 bar (exactly $10^5$ Pa), as is used today.

If we know the general expression of the Gibbs energy function with pressure, we can write:

$$\frac{\partial \ln K_r^{(P)}}{\partial P^0} = \frac{1}{\mathrm{R}T}\frac{\partial \Delta_r g^0}{\partial P^0} = -\frac{\Delta_r v_0^0}{\mathrm{R}T} \qquad [3.36]$$

However, for perfect gases, we have:

$$\Delta_\mathrm{r} v_0^0 = \frac{\mathrm{R}T}{P^0}\sum_i v_i \qquad [3.37]$$

Hence the variations of the constant with the reference pressure:

$$\frac{\partial \ln K_r^{(P)}}{\partial P^0} = -\frac{\sum_i v_i}{P^0} \qquad [3.38]$$

Between two reference pressures $P_{(1)}^0$ and $P_{(2)}^0$, by integration we find:

$$\frac{K_{r(2)}^{(P_2^0)}}{K_{r(1)}^{(P_2^0)}} = \left(\frac{P_{(1)}^0}{P_{(2)}^0}\right)^{\sum_k v_k} \qquad [3.39]$$

In the example noted below, of the change of reference for the data tables, the correction is generally very slight and is often ignored.

NOTE 3.7.– Certain transformations of pure phases or solutions can be symbolically represented by a balance equation, which is generally very simple. We can very easily extend the concept of the equilibrium constant and law of mass action.

For the liquid–vapor balance, we can write a balance equation:

$$(\mathrm{A}) = \{\mathrm{A}\} \qquad [3R.1]$$

The application of the law of mass action (adopting the hypothesis of a low vapor pressure which means that fugacity and pressure can be regarded as the same thing) becomes:

$$\frac{P^{(\mathrm{vap})}}{P^0} = K^{(P)(\mathrm{vap})} \qquad [3.40]$$

We can see that the equilibrium constant is the vaporization constant $K^{(P)(\mathrm{vap})}$. As the reference pressure is often 1 bar, the vaporization constant is the saturating vapor pressure at the temperature in question: $P_0^{0(\mathrm{vap})}$.

Thus, by the same type of reasoning, we would find that the constants of vaporization, demixing and sharing of a solute between two solvents are equilibrium constants.

### 3.1.8. *Dissociative dissolution of a gas in a solid*

To conclude this section on equilibrium constants, we are going to discuss the case where the superposition of a phase equilibrium is from dissociation, in the case of the dissociative dissolution of a gas in a solid.

Let us envisage the dissolution of hydrogen in palladium. The experimental results were interpreted by the dissociation of the hydrogen molecule at the moment of dissolution and insertion of hydrogen atoms in interstitial sites <s>

in the solid. In fact, the inserted atoms of H and the free sites form a solid solution and we shall write the balance equation in the form:

2<<s>> + $\{H_2\}$ = 2<<H>> [3R.2]

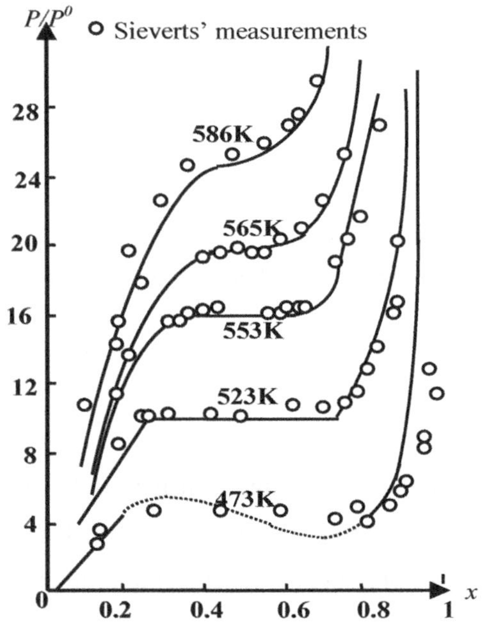


**Figure 3.1.** *Isothermal of pressure to composition for dissolution of hydrogen in palladium. Curves calculated by Larcher; [FOW 49]*

By taking convention (I) as the reference of the solution which has two components, the localized hydrogen atoms and the free sites, the law of mass action is written as:

$$\left[\frac{\gamma_{\mathrm{H}}^{(\mathrm{I})} x}{\gamma_{\mathrm{s}}^{(\mathrm{I})}(1-x)}\right]^2 \frac{P^0}{P_{\mathrm{H}_2}} = K_2^{(P)} \qquad [3.41]$$

By choosing the model of a strictly-regular solution for the solid solution, we obtain the results shown in Figure 3.1, where we show the hydrogen pressure at equilibrium as a function of its atomic fraction in the solid. Note that we have a very good coincidence between the curves given by the model and the experimental points. These results are as satisfactory for zones with a single solid phase as for zones of demixing of the solid solution which exhibit a platform.

## 3.2. Graphical representations of equilibria – pole diagrams

A certain number of graphical representations are used to visualize the results of a chemical equilibrium. Some of these graphs superpose several equilibria, and others also add the results of physical equilibria of state change. In this section, we examine one series of such diagrams; others are introduced when we look at particular systems such as equilibria in aqueous solutions.[1]

The pole diagram is a helpful tool to represent the progression of a reaction and compare reactions within the same family. It is generally plotted in convention (I), (pure-substance reference).

### 3.2.1. *Principle of the pole diagram*

The pole diagram (see Figure 3.2) is a representation of the variations in the affinity as a function of the quantity $\mathrm{R}\ln Q^{(\mathrm{I})}$ for a reaction at a certain temperature $T$.

In Figure 3.2, P denotes the point with coordinates $(\Delta_r s^0, -\Delta_r h^0)$, at standard entropy and enthalpy – i.e. in reference (I) – associated with the reaction under study. These values may be positive or negative (in the case of Figure 3.2, we have chosen negative values for $\Delta_r s^0$ and $\Delta_r h^0$). P is called the pole of the reaction. It characterizes the physical states (states of aggregation) of the components at the chosen temperature $T$. The straight line which, for the reaction at hand at temperature $T$, represents the variations in affinity as a function of $\mathrm{R}\ln Q^{(\mathrm{I})}$, passes through that point P. it also passes through the point I, which represents the real experimental conditions at the initial state of the reaction. The ordinate of that point I is the initial affinity, its abscissa value is $\mathrm{R}\ln Q_0^{(\mathrm{I})}$, and $Q_0^{(\mathrm{I})}$ is the initial value of $Q^{(\mathrm{I})}$. Point B on the line, on the axis of the affinities, represents the

1 See Volume 6 of this Set of books, *Electrolytes and Electrochemical Thermodynamics*.

standard affinity. Point E, whose ordinate is null, represents the state of equilibrium of the reaction because at equilibrium, the affinity is null, its abscissa value is obviously $R\ln K^{(\mathrm{I})}$. Thus, the segment IE gives the affinity over the course of the reaction.

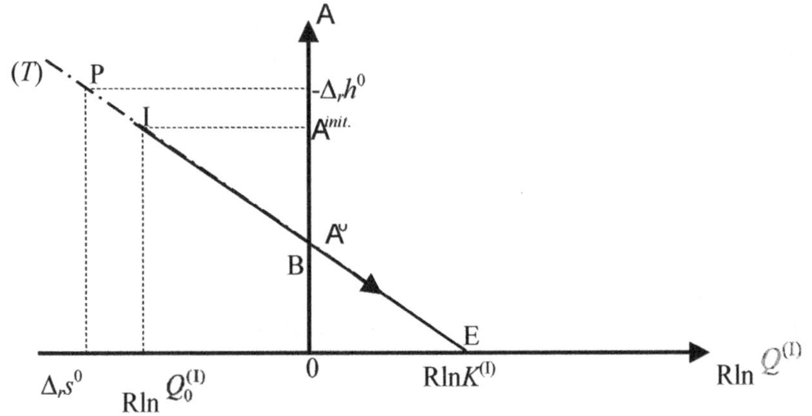


**Figure 3.2.** *Pole diagram of a reaction*

### 3.2.2. *Influence of a temperature change on a pole diagram*

If we perform the reaction at a temperature $T'$ which is higher than $T$, two cases may arise:

1) If, at the new temperature, the states of aggregation of the components are the same as at temperature $T$, the new line again passes through the pole P, the line has simply rotated around P (see Figure 3.3). If the initial concentrations are the same as in the previous case, this means that we start with the same value of the reaction quotient, and thus we start from point I', with the same abscissa value as I.

This time, we reach point E', with abscissa $R\ln K'^{(\mathrm{I})}$. The reaction now runs along the segment I'E'.

2) If, at the new temperature $T'$, the state of aggregation of a component is no longer the same as at $T$ in the standard state, meaning that this component has undergone a phase transformation (written as $t$) between the temperatures $T$ and $T'$. We therefore need to define a new pole P'

(Figure 3.4). The reaction now runs along the segment I'E'. Consider the line PP'; its slope, which is always negative, gives the temperature of the phase transformation of the component in question. Indeed, the slope of line PP' is:

$$a = \frac{\Delta_r h'^0 - \Delta_r h^0}{\Delta_r s^0 - \Delta_r s'^0} \qquad [3.42]$$

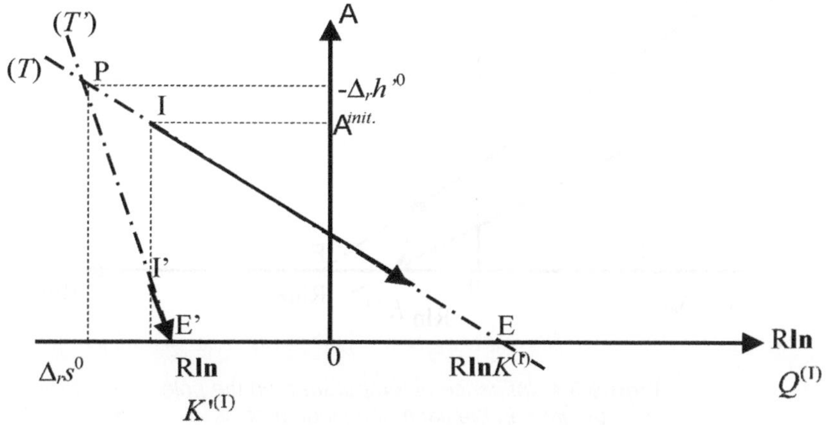


**Figure 3.3.** *Influence of temperature with conservation of the same pole*

However, the difference $\Delta_r h'^0$-$\Delta_r h^0$ represents the enthalpy $\Delta_t h^0$ accompanying the phase change of the component in question between P and P'. Similarly, the difference $\Delta_r s^0$-$\Delta_r s'^0$ represents the opposite of entropy $\Delta_t s^0$ associated with the same transformation. The slope of the straight line PP' is therefore given by the opposite of the ratio of the enthalpy of phase change of the compound involved to the corresponding entropy:

$$a = -\frac{\Delta_t h^0}{\Delta_t s^0} \qquad [3.43]$$

However, at the temperature of that phase transformation, we have:

$$\Delta_t g^0 = \Delta_t h^0 - T_t \Delta_t s^0 = 0 \qquad [3.44]$$

Thus, the slope of line PP' becomes:

$$a = -T_t \qquad [3.45]$$

This result confirms our statement.

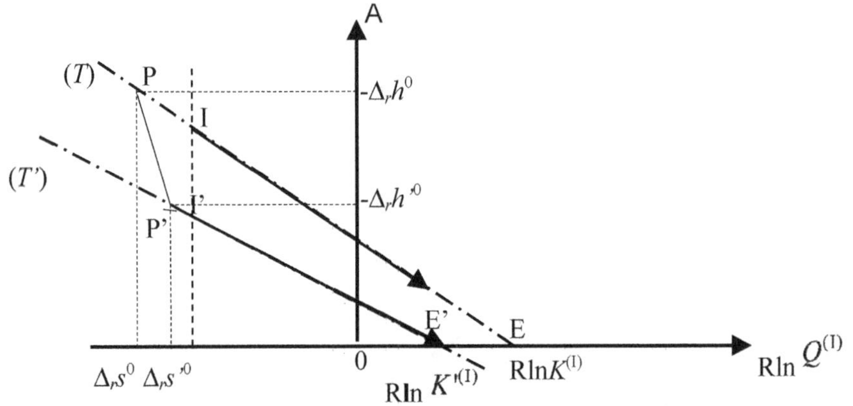


**Figure 3.4.** *Influence of temperature on the pole diagram in the case of changing poles*

### 3.2.3. *Pole diagrams of two reactions in the same family*

It is often interesting to use the same system of axes to plot the pole diagrams of two reactions involving similar systems. We shall now examine two examples.

Let us first look at the reactions of reduction of the same metal oxide MO, by hydrogen on the one hand, and by carbon monoxide on the other.

The two reactions are written:

$$MO + H_2 = M + H_2O \qquad [3R.3]$$

$$MO + CO = M + CO_2 \qquad [3R.4]$$

At the same temperature, each of the reactions is represented by a pole – P and P' respectively. In the same way as in the previous section, we can demonstrate that the slope of line PP' (which is necessarily negative) gives the temperature of inversion of the reaction:

$$CO_2 + H_2 = CO + H_2O \quad [3R.5]$$

This is the temperature for which, at equilibrium, we have the relation:

$$\frac{P_{CO}}{P_{CO_2}} = \frac{P_{H_2O}}{P_{H_2}} \quad [3.46]$$

Similarly, let us now compare the pole diagrams of reduction of two metal oxides, MO and M'O by hydrogen, as follows:

$$MO + H_2 = M + H_2O \quad [3R.6]$$

$$M'O + H_2 = M' + H_2O \quad [3R.7]$$

The straight line joining the two poles P and P' of those two reactions at the same temperature gives the equilibrium temperature of the reaction between solids:

$$MO + M' = M + OM' \quad [3R.8]$$

## 3.3. Representation of the evolution of an equilibrium with the temperature

As temperature is an important variable of chemical equilibrium, users have attempted to represent the evolution of a chemical equilibrium with changing temperature. Two methods are discussed below.

### 3.3.1. *Diagram in van 't Hoff coordinates*

The first mode of representation is based on relation [3.28], applied in convention (I). The method involves representing, in standard pressure conditions (in practice at the pressure of 1 bar), the logarithm of the equilibrium constant as a function of the inverse of temperature. This representation gives us practically a straight line, because the standard enthalpy $\Delta_r h^0$ and standard entropy $\Delta_r s^0$ of the reaction are practically independent of the temperature. Hence, the slope of that line may be

$-\Delta_r h^0 / \mathrm{R}$, and its ordinate at the origin may be $\Delta_r s^0 / \mathrm{R}$ by virtue of relation [3.28].

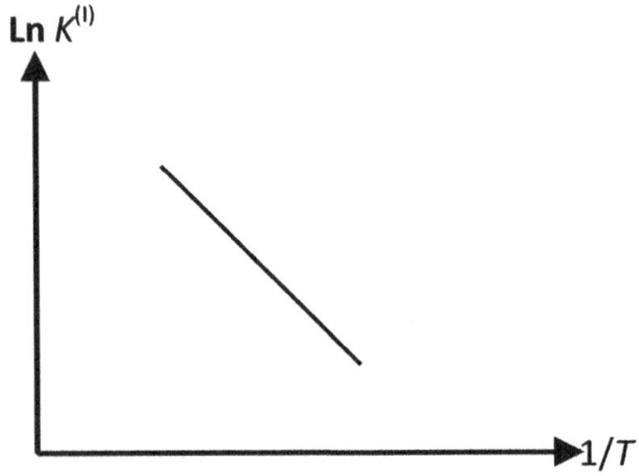


**Figure 3.5.** *Representation of the evolution of an equilibrium with temperature as a van 't Hoff diagram*

Figure 3.5 shows such a line in the case of an endothermic reaction (positive reaction enthalpy).

### 3.3.2. *Ellingham diagrams*

The second mode of representation of the evolution of an equilibrium with the temperature is the generalized Ellingham diagram, which we shall now examine in detail.

#### 3.3.2.1. *Ellingham representation*

Consider the context of the pure-substance reference (I). The principle of that diagram is, at standard pressure, to plot the standard Gibbs energy $\Delta_r g^0$ for the reaction in the plane $[T,\ \mathrm{R}T\ln Q^{(\mathrm{I})}]$ (Figure 3.6(a)). Using relation [3.44], we can see that if the standard enthalpy $\Delta_r h^0$ and standard entropy $\Delta_r s^0$ of the reaction are practically independent of temperature (these are the so-called Ellingham approximations), the representative curve is a segment of straight line whose slope is the opposite of the standard entropy $\Delta_r s^0$ and the intercept is the standard enthalpy $\Delta_r h^0$.

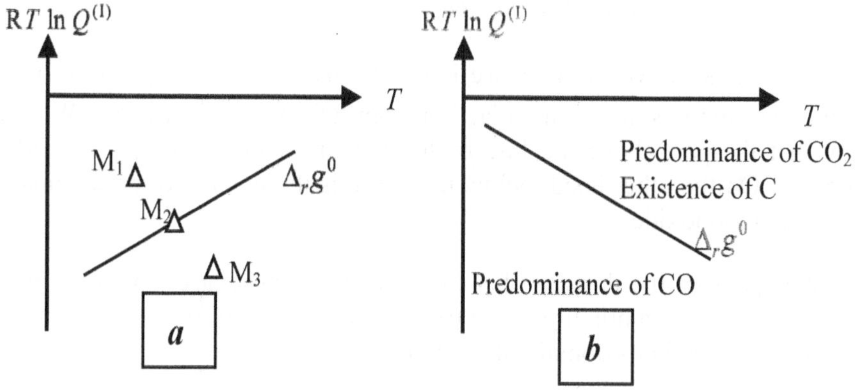


**Figure 3.6.** *a) Generalized Ellingham diagram; b) Ellingham diagram for the $CO_2$ +C=2CO equilibrium*

– if we choose a point above the curve such as the point $M_1$ in Figure 3.6(a). We have $RT \ln Q^{(l)} > \Delta_r g^0$, so by virtue of relation [3.1], the affinity is positive and the reaction takes place from left to right;

– if we choose a point below the curve such as the point $M_3$ in Figure 3.6(a). We have $RT \ln Q^{(l)} < \Delta_r g^0$, so by virtue of relation [3.1], the affinity is negative, and the reaction is impossible. Hence, we see the opposite reaction, which takes place from right to left;

– if we choose a point on the curve such as point $M_2$ in Figure 3.6(a). We have $RT \ln Q^{(l)} = \Delta_r G^0$, so by virtue of relation [3.1], the affinity is null and the reaction is at thermodynamic equilibrium. Thus, the curve illustrates the evolution of the equilibrium conditions with changing temperature.

### 3.3.2.2. *Domains of predominance, domains of existence*

Consider the equilibrium:

$$CO_2 + C = 2CO \qquad [3R.9]$$

Let us plot the Ellingham diagram for that equilibrium. As the reaction takes place from left to right with an increase in the number of gaseous molecules, the standard entropy will be positive, and therefore the slope of the line will be negative (Figure 3.6(b)).

If we look above the line, we have $RT \ln Q^{(I)} > \Delta_r g^0$. By virtue of the same reasoning as above, the reaction from left to right is favored. We then say that the part above the line is the domain of predominance of carbon dioxide. On the other hand, below the curve is the domain of predominance of carbon monoxide.

We speak of the domain of predominance because at every point of the diagram, we have a mixture of the two gases, but each of them is present in a greater quantity in its domain of predominance.

On the other hand, for carbon, above the curve the system contains carbon, whilst below the curve the carbon will have completely disappeared, so we say that above the curve is the domain of existence of carbon.

More generally, for the components belonging to a phase with multiple components (mixtures of gases or solutions), the Ellingham curve delimits domains of predominance, whereas for components in a pure phase, the Ellingham curve delimits domains of existence.

#### 3.3.2.3. *Ellingham diagram for metal–oxygen–oxide systems*

At the heart of it, the Ellingham diagrams were constructed and used for the stability of metal oxides. In order to facilitate comparisons and combinations of reactions, we tend to write the balance equations of those reactions involving only one oxygen molecule, as in the following examples:

$$C + O_2 = CO_2 \quad [3R.10]$$

$$2C + O_2 = 2CO \quad [3R.11]$$

$$4/3Al + O_2 = 2/3\ Al_2O_3 \quad [3R.12]$$

In these conditions, if $P$ is the oxygen pressure at equilibrium, the line represents the quantity $\Delta_1 g^0 = RT \log P$ as a function of the temperature.

The slope of the line is positive or negative depending on the sign of the standard entropy of the reaction, and in practice, depending on the creation or consumption of gas molecules in the reaction. Thus, in reaction [3R.10], with the number of gaseous moles being the same on both sides of the equation, the entropy and therefore the slope of the Ellingham line is practically non-existent. In reaction [3R.11], there is an increase in the number of gaseous moles because of the reaction, and therefore an increase in entropy, resulting in a negative slope. For reaction [3R.12], there is a decrease in the number of gaseous moles, so the standard entropy decreases and the slope of the Ellingham line is positive.

### 3.3.2.4. *Oxygen pressure at equilibrium at a given temperature in a metal-oxide Ellingham diagram*

Consider the Ellingham line relative to a reaction $r_1$ (Figure 3.7(a)). Let Ω denote the origin of the axes – i.e. the point with the coordinates (-273°C, 0). We place ourselves at a temperature *T*, the abscissa of a point H and let $M_1$ be the point of intersection of the vertical issuing from H and the Ellingham line of the reaction $r_1$. At another given temperature *T'* (the abscissa of point H'), we again place a vertical axis y'y. The line $\Omega M_1$ cuts that axis at point $M'_1$. We can show that all the points on the line $\Omega M'_1$ correspond to the same oxygen pressure.

In the two homothetic triangles $\Omega H M_1$ and $\Omega H' M'_1$, we can write:

$$\frac{\overline{HM_1}}{\overline{H'M'_1}} = \frac{\overline{\Omega H}}{\overline{\Omega H'}} \qquad [3.47a]$$

so

$$\frac{\Delta_1 g^0}{\Delta_1 g'^0} = \frac{RT \log P_1}{RT' \log P'_1} = \frac{T}{T'} \qquad [3.47b]$$

From this we deduce that the pressure at point $M'_1$ is the same as at point $M_1$, and therefore that, whatever the position of the y'y axis, we have:

$$\log P_1 = \log P'_1 \qquad [3.48]$$

Now consider a second reaction $r_2$ and let $\Delta_2 g^0$ be the standard Gibbs energy associated with the temperature *T*, which is *k* times higher than that of reaction $r_1$ at the same temperature, and thus:

$$\Delta_2 g^0 = k\,\Delta_1 g^0 \qquad [3.49]$$

Hence:

$$\overline{\mathrm{HM}_2} = k\overline{\mathrm{HM}_1} \qquad [3.49a]$$

We can deduce from this that the two equilibrium pressures are such that:

$$\mathrm{R}T \ln P_2 = k\mathrm{R}T \ln P_1 \qquad [3.50]$$

Thus:

$$\ln P_2 = k \ln P_1 \qquad [3.51]$$

The y'y axis, therefore, is an axis of a linear scale of the logarithms of the partial pressures, and thus a logarithmic scale of those pressures. This axis is often divided into bars, in accordance with the logarithmic scale.

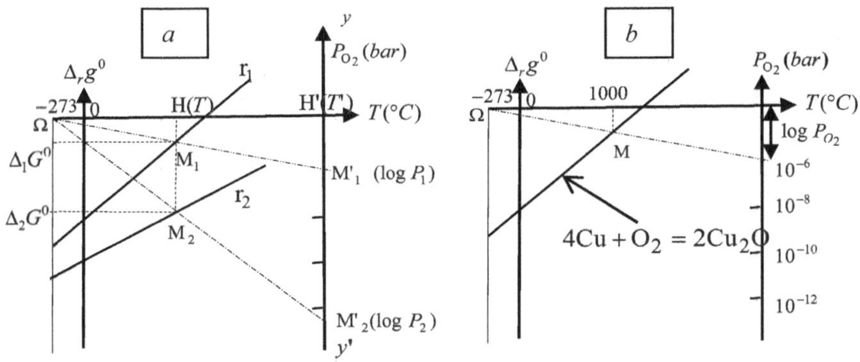


**Figure 3.7.** *Ellingham diagram and oxygen pressure at equilibrium*

When we know those properties, it is easy to determine the oxygen pressure at equilibrium at a certain temperature, using the graph. In order to do so (see Figure 3.7b), consider the equilibrium between copper and copper oxide; we determine the point M (at 1000°C on the figure) and the line ΩM, which we stretch to that y'y axis, on which we read the equilibrium pressure between copper and copper oxide at 1000°C, which is around $10^{-6}$ bars.

### 3.3.2.5. *Effects of changes of state of a component of the reaction in a metal–oxide Ellingham diagram*

The slope of the Ellingham line is modified when one of the components of the reaction undergoes a change of physical state at a certain temperature, in the wake of the entropy accompanying that change of state. We know that the entropy of state change is positive in transformations from the solid state to the liquid state and from the liquid state to the gaseous state, meaning when the degree of order decreases.

If a component on the left-hand side of the reaction – say, aluminum in reaction [3R.12] – undergoes fusion or volatilization, the standard entropy of the reaction is decreased by the entropy of state change, and therefore there is a break in the slope of the Ellingham line ($-\Delta_r s^0$), which sharply increases at the state-change temperature.

On the other hand, if a component on the right-hand side of the reaction undergoes fusion or volatilization, the standard entropy of the reaction is increased by the entropy of state change, and thus the slope of the Ellingham line ($-\Delta_r s^0$) lessens.

Figure 3.8 illustrates the consequences of those two types of state changes.

However, a polymorphic state change to the solid state, which is generally accompanied by a negligible entropy of state change, has practically no effect on the slope of the line.

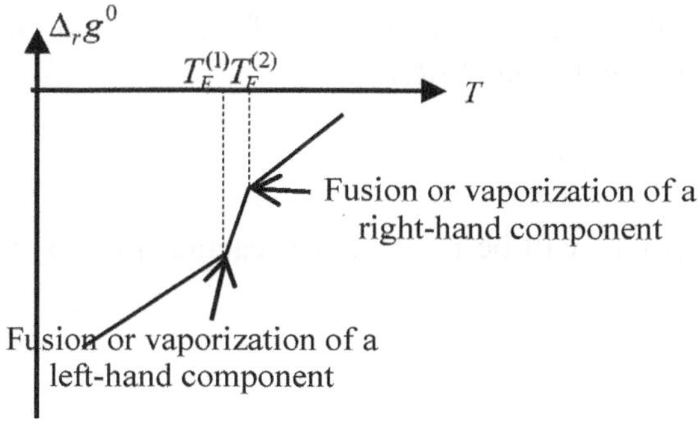


**Figure 3.8.** *Consequences of state changes of a component on an Ellingham diagram*

### 3.3.2.6. *Redox reactions and inversion temperature in a metal-oxide Ellingham diagram*

Ellingham diagrams enable us to discover, at a given temperature, the direction of the reaction between a metal and an oxide of a different metal. By way of example, let us consider the redox reaction between silicon and chrome oxide, which would be written as:

$$\mathrm{Si} + 2/3\mathrm{Cr_2O_3} = \mathrm{SiO_2} + 4/3\mathrm{Cr} \qquad [3R.13]$$

To this end, let us consider the following two reactions of chromium and silicon with oxygen (reactions [3R.14] and [3R.15]), each represented by an Ellingham line (Figure 3.9(a)):

$$4/3\mathrm{Cr} + \mathrm{O_2} = 2/3\mathrm{Cr_2O_3} \qquad [3R.14]$$

$$\mathrm{Si} + \mathrm{O_2} = \mathrm{SiO_2} \qquad [3R.15]$$

We can see that the balance equation for reaction [3R.13] is the difference between the two balance equations for reactions [3R.14] and [3R.15], so according to relation [3.22], its affinity will be:

$$\mathscr{A}_{13} = \mathscr{A}_{15} - \mathscr{A}_{14} \qquad [3.52]$$

Thus, as a function of the standard Gibbs energies for the two reactions [3R.14] and [3R.15]:

$$\mathscr{A}_{13} = (-\Delta_{15}g^\circ + RT\ln P_{O_2}) - (-\Delta_{14}g^\circ + RT\ln P_{O_2}) \qquad [3.53]$$

Hence, this affinity is only a function of the standard Gibbs energies of the two reactions [3R.14] and [3R.15]:

$$\mathscr{A}_{13} = \Delta_{14}g^\circ - \Delta_{15}g^\circ \qquad [3.54]$$

Thus, this affinity will be positive, so reaction [3R.13] is possible from left to right if:

$$\Delta_{14}g^\circ > \Delta_{15}g^\circ \qquad [3.55]$$

This means that the Ellingham line of reaction [3R.15] must be placed below that of reaction [3R.14], as shown by Figure 3.9(a).

Thus, generally speaking, a metal is capable of reducing an oxide if its Ellingham line is situated below that of the oxide.

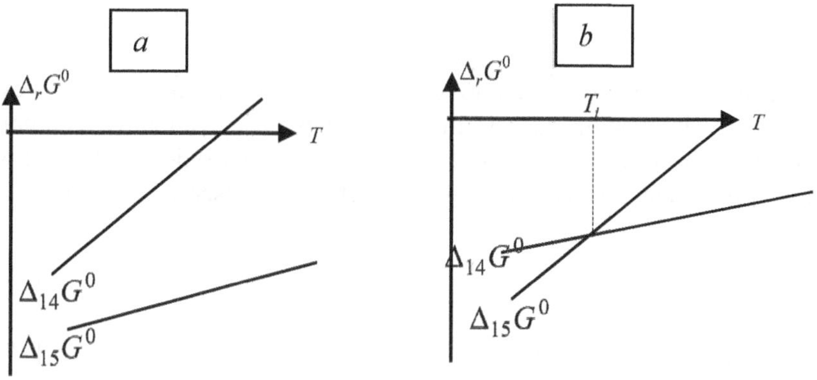


**Figure 3.9.** *a) Possibilities of reactions between metal–oxide couples; b) case where the Ellingham lines intersect one another*

In our chosen example of silicon and chromium, the two Ellingham lines relating to the two reactions did not intersect in the temperature domain in question. Now let us look at what happens if those two curves intersect at a temperature $T_i$. For that purpose, we choose the two reactions [3R.14] and [3R.16], whose two Ellingham lines are secants (Figure 3.0(b)):

$$4Na + O_2 = 2Na_2O \quad [3R.16]$$

The potential reaction between solids will be:

$$4Na + 2/3Cr_2O3 = 4/3Cr + 2Na_2O \quad [3R.17]$$

If the temperature is lower than $T_i$, the straight line pertaining to the reaction [3R.16] is situated below that relative to the reaction [3R.14] so sodium will reduce chromium oxide; the reaction [3R.17] takes place spontaneously from left to right. On the other hand, for temperatures higher than $T_i$, the straight line relative to the reaction [3R.14] is situated below that relating to the reaction [3R.16], and thus chromium will reduce sodium oxide, with reaction [3R.17] taking place spontaneously from right to left.

Temperature $T_i$ is known as the temperature of inversion of reaction [3R.17].

#### 3.3.2.7. *Ellingham diagram for a metal with multiple oxides – diagram for iron*

Certain metals are capable of forming several different oxides. A particularly interesting case is that of iron oxides, with three compounds elementary iron, iron monoxide FeO, and tri-iron tetroxide $Fe_3O_4$. Between those components, we can write three reactions represented by an Ellingham line:

$$2Fe + O_2 = 2FeO \qquad [3R.18]$$

$$6FeO + O_2 = 2Fe_3O_4 \qquad [3R.19]$$

$$3/2Fe + O_2 = 1/2Fe_3O_4 \qquad [3R.20]$$

As these three reactions are not independent, their standard Gibbs energies are linked to one another by the relation:

$$\Delta_{20}g^0 = \frac{1}{4}\Delta_{19}g^0 + \frac{3}{4}\Delta_{18}g^0 \qquad [3.56]$$

Figure 3.10(a) gives the representation of the three Ellingham lines, which intersect at temperature $T_t = 825$ K.

The straight line pertaining to equilibrium [3R.20] is situated between those representing the equilibria [3R.18] and [3R.19].

Let us first place ourselves at a temperature lower than 825 K. By consideration of the stable components in relation to each curve, we can easily show that there is only compatibility between the curves if we conserve that relating to the equilibrium [3R.20] (Figure 3.10(b)), and therefore two zones of stability: one for metal iron (bottom) and the other for tetroxide (bottom).

On the other hand, below the temperature of 825 K, it is easy to demonstrate that we can only conserve (Figure 3.10(b)) the curves relative to the equilibria [3R.20] and [3R.16], and therefore ultimately the three zones of stability illustrated in Figure 3.10(c).

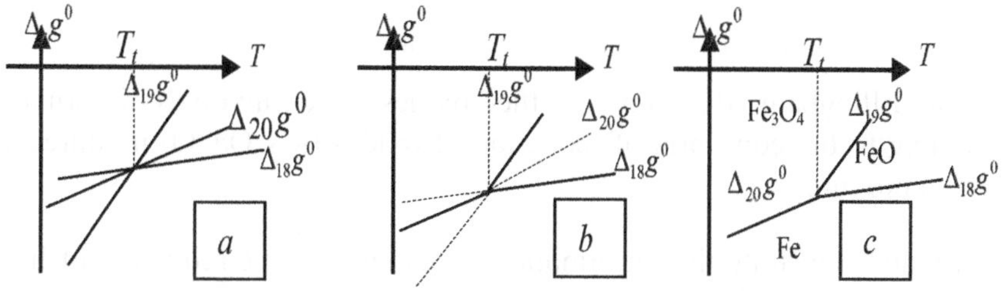


**Figure 3.10.** *Diagrams of the stability of iron oxides*

The temperature of 825 K is called the transition temperature.

NOTE 3.8.– there is a fourth possible reaction between the same three components, which is the dismutation of monoxide, as follows:

$$4FeO = Fe + Fe_3O_4 \quad [3R.21]$$

For this reaction, the standard Gibbs energy is easily calculated on the basis of the relation:

$$\Delta_{21}g^0 = \frac{1}{2}\left(\Delta_{19}g^0 - \Delta_{18}g^0\right) \quad [3.57]$$

Thus, depending on the temperature, we distinguish:

– if $T = T_t$, we have: $\mathscr{A}_{21} = -\Delta_{21}g^0 = 0$, and reaction [3R.21] is at equilibrium;

– if $T > T_t$, we have: $\mathscr{A}_{21} = -\Delta_{21}g^0 < 0$, the oxide FeO cannot exist, so and reaction [3R.21] is shifted entirely to the right;

– if $T < T_t$, we have: $\mathscr{A}_{21} = -\Delta_{21}g^0 > 0$, then the metal iron cannot exist, so the reaction [3R.21] is shifted entirely to the left.

NOTE 3.9.– we note that the above reasoning hardly applies at all to iron sesquioxide $Fe_2O_3$. Indeed, that compound is not stable in the presence of iron or of one of the other oxides, and does not reach equilibrium.

### 3.3.2.8. *Extension of the metal–oxide Ellingham diagrams to redox by gaseous mixtures of oxidizing/reducing agents*

The Ellingham diagram of the oxides also involves information pertaining to the equilibria of reduction of oxides by $CO/CO_2$ mixtures and $H_2/H_2O$ mixtures.

We shall illustrate that information by examining the reduction of liquid zinc oxide at 700°C by the $CO/CO_2$ mixture.

For that, we use the data relative to the equilibria:

$$2CO + O_2 = 2CO_2 \qquad [3R.22]$$

$$2Zn_{liq} + O_2 = 2ZnO \qquad [3R.23]$$

The reaction under study is given by the balance equation:

$$CO + ZnO = CO_2 + Zn \qquad [3R.24]$$

We can see that, between the three balance reactions, we have the relation:

$$[3R.22] - [3R.23] = 2[3R.24] \qquad [3.58]$$

At equilibrium, the affinity of reaction [3R.24] can be written in the form:

$$\mathcal{A}_{24} = \mathcal{A}_{24}^{0} - RT\ln\left(\frac{P_{CO_2}}{P_{CO}}\right)_{equ} = 0 \qquad [3.59]$$

In view of relation [3.58], the standard affinity is given by:

$$\mathcal{A}_{24}^{0} = -\Delta_{24}g^{0} = -\frac{1}{2}\left(\Delta_{22}g^{0} - \Delta_{23}g^{0}\right) \qquad [3.60]$$

By taking account of equations [3.59] and [3.60], we obtain:

$$\Delta_{22}g^{0} - 4.6RT\log\left(\frac{P_{CO_2}}{P_{CO}}\right)_{equ} = \Delta_{23}g^{0} \qquad [3.61]$$

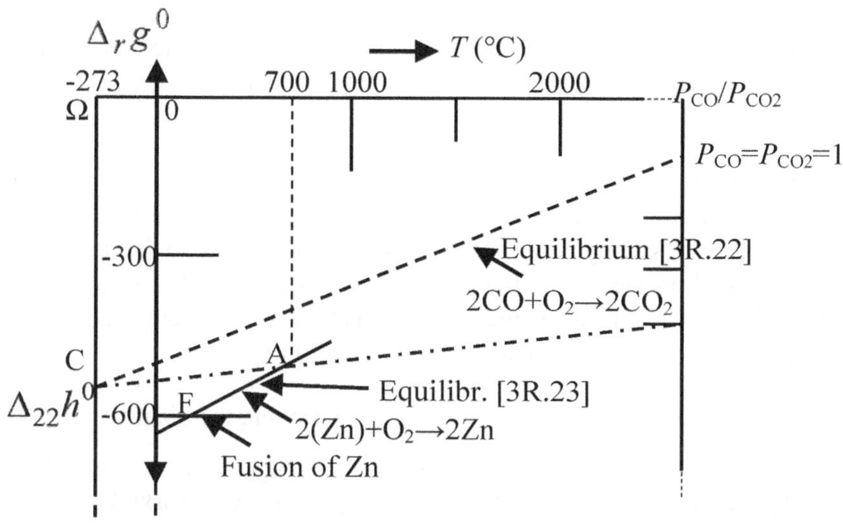


**Figure 3.11.** *Reduction of zinc oxide by the CO/CO$_2$ mixture*

Figure 3.11 shows:

– the origin Ω of the axes (-273°C, 0 kJ);

– the straight line relative to equilibrium [3R.22] in standard conditions. This straight line begins at point C of the coordinates (-273°C, $\Delta_{22}g^0(\text{at } 0\ K) = \Delta_{22}h^0 =$ 552 kJ), and finishes at a scale in a decimal logarithm of the ratio $P_{CO2}/P_{CO}$ at the value which corresponds to 1 bar for each of the partial pressures (standard conditions);

– the straight line relative to equilibrium [3R.23] in standard conditions, for zinc in the liquid state – i.e. beyond the fusion point F of the metal;

– point A, obtained by the intersection of the above line and the vertical at 700°C;

– the straight line with the equation:

$$Y = \Delta_{22}g^0 - 4.6\text{R}T\log\left(\frac{P_{CO_2}}{P_{CO}}\right)_{equ} \qquad [3.62]$$

We can see from the figure that, in standard conditions, zinc oxide cannot be reduced by carbon monoxide, because the Ellingham line for equilibrium [3R.23] is situated, at 700°C, below that relative to equilibrium [3R.24].

The straight line given by equation [3.52] intersects the straight line relative to equilibrium [3R.23] at point A, according to relation [3.51]. Thus, the intersection of this straight line with a horizontal placed above the diagram or a vertical placed on the side if the diagram defines a scale $P_{CO2}/P_{CO}$, graduated from $10^{-8}$ to $10^{10}$. Thus, extending line CA enables us to directly read the value of that ratio at equilibrium of reaction [3R.24] at 700°C.

Similarly, for the reduction of oxides by hydrogen, the diagram contains:

– a line corresponding to the oxidation of hydrogen, which begins from a point H with the coordinates (-273°C, $\Delta_r g^0(\grave{a}0\ K) = \Delta_r h^0 = 480$ kJ);

– a scale of the ratios of pressures $P_{H2O}/P_{H2}$.

### 3.3.2.9. *Complete Ellingham diagram for oxide–metal systems*

Having chosen to write all the reactions with a gaseous molecule on the left-hand side, we can plot the Ellingham lines for the different reactions on the same graph.

Figure 3.12 shows plots all of the oxides diagram on one graph. Here are see the graduation of the standard Gibbs energies on the left-hand vertical axis, the point of origin Ω, and on the right-hand vertical axis, the logarithmic scale in oxygen pressure. The figure also shows points C and H. In order to make it easier to plot, we have not shown the graduated axes in $P_{H_2O} / P_{H_2}$ and $P_{CO_2} / P_{CO}$.

We can easily use the diagram to find slope ruptures due to changes of the physical state, which are identified, by the letters F, B, S and T (fusion, boiling, sublimation and polymorphic transformation respectively).

Also note the straight line of oxidation of the carbon in carbon monoxide, which has a negative slope because of the increase in the number of gaseous molecules during the course of the reaction, which gives highly-positive entropy. Similar diagrams exist for nitrides and carbides [DES 10].

### 3.3.2.10. *Ellingham diagram and solutions solids*

In the case where the solids involved in the reaction belong to solid solutions, their activities play a part in the expression of the affinity which is cancelled out at equilibrium, for a metal oxide MO, in accordance with:

$$\mathscr{A} = \mathscr{A}^0 - RT \ln Q^{(1)} = -\Delta_r g^0 + RT \ln\left(\frac{P_{O_2}}{P^0} x \frac{a_M^{(1)}}{a_{MO}^{(1)}}\right) = 0 \qquad [3.63]$$

At equilibrium, we obtain the expression of the standard Gibbs energy:

$$\Delta_r g^0 = RT \ln\left(\frac{P_{O_2}}{P^0} \times \frac{a_{\text{M}}^{(1)}}{a_{\text{MO}}^{(1)}}\right) \qquad [3.64]$$

Thus, we can use the Ellingham diagram, but the scale of the oxygen pressures in fact gives us the product $\frac{P_{O_2}}{P^0} \times \frac{a_{\text{M}}^{(1)}}{a_{\text{MO}}^{(1)}}$.

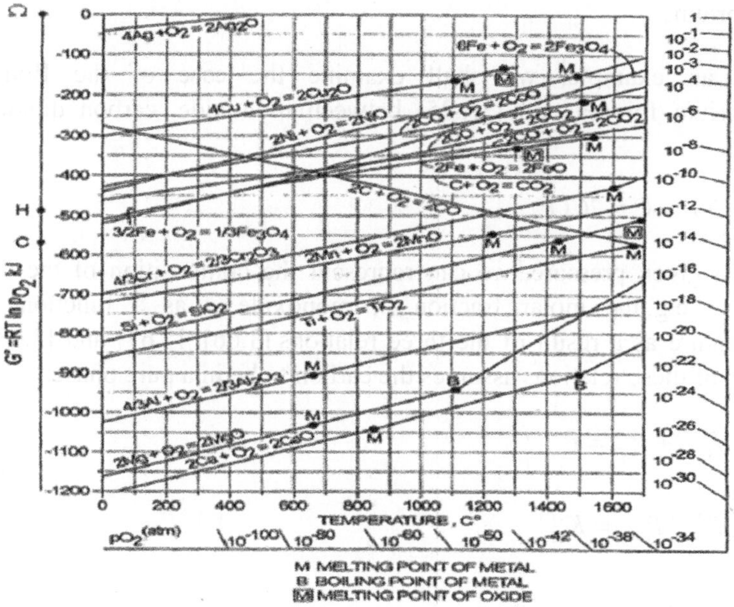


**Figure 3.12.** *Ellingham diagrams for the oxidations by gaseous oxygen (www.google. com/patents/EP1218556B1?cl=fr,2004)*

If the solutions are perfect, the activities are replaced by the molar fractions of the components in their respective solutions, and we have:

$$\Delta_r g^0 = RT \ln\left(\frac{P_{O_2}}{P^0} \times \frac{x_M}{x_{MO}}\right) \quad [3.65]$$

The oxygen pressure scale then gives us the product $\frac{P_{O_2}}{P^0} \times \frac{x_M}{x_{MO}}$.

## 3.4. Binary diagrams for chemical equilibrium

For the equilibria which involve a single phase with two components (whether pure-phase or not), gaseous or liquid solution or solid, we use diagrams with the same coordinates as the binary phase diagrams, meaning that they give the temperature as a function of the composition of the two-component phase (or *vice versa*). These are binary diagrams for chemical equilibrium.

As an example, we shall examine the case of the Boudouard equilibrium for reaction [3R.25] between monoxide, carbon dioxide and carbon.

$$CO_2 + C = 2CO \quad [3R.25]$$

At a given pressure, we can represent the composition of the gaseous phase – e.g. the molar fraction of monoxide – as a function of the temperature as a result of the three relations [3.66], [3.67] and [3.68]. The second of these relations assumes the carbon to be in a pure phase:

$$P_{CO_2} + P_{CO} = P \quad [3.66]$$

$$P_{CO}^2 / P_{CO_2} = K_{25}^{(P)} \quad [3.67]$$

$$\frac{d \ln K_{25}^{(P)}}{dT} = \frac{\Delta_{25} h^0}{RT^2} \quad [3.68]$$

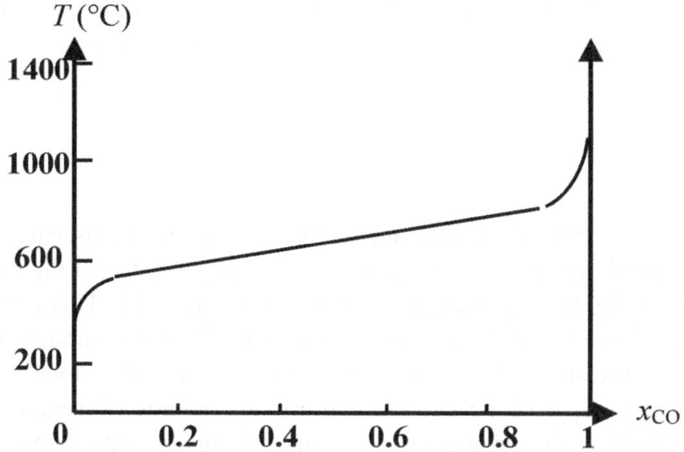


**Figure 3.13.** *Boudouard diagram for a total pressure of 1 atmosphere*

Figure 3.13 shows the binary diagram of that reaction for a total pressure of 1 bar. If the carbon is not of pure phase – e.g. if we have the same equilibrium in the presence of steel (solution of carbon in iron) – then relation [3.68] is replaced by:

$$P_{CO}^2 / a_C^{(1)} P_{CO_2} = K_P \quad [3.69]$$

This formula involves the carbon activity in convention (I). Therefore, the diagram in Figure 3.13 will be plotted, with the total pressure and carbon activity being constant.

## 3.5. Ternary diagrams of chemical equilibria

The purpose of the ternary diagram is to enable us to quickly read the composition at equilibrium at a certain temperature and a certain pressure, knowing the initial conditions, for reactions involving a single phase with three components (mixture of three gases or condensed solution with components). As the reaction may also involve pure phases – e.g. solids – we shall write it very generally with its three components in the polycomponent phase $A_1$, $A_2$ and $A_3$, in the form:

$$\text{solids} + \beta_1 A_1 + \beta_2 A_2 = \beta_3 A_3 + \text{solids} \quad [3R.26]$$

Coefficient $\beta_2$ may have a value of 0, in which case the equilibrium is reached in the presence of an inert component of the solution – e.g. an inert gas $A_2$.

### 3.5.1. *Mode of representation*

The composition of the mixture of the three components can, as in the case of phase diagrams for ternary systems, be represented in the equilateral triangle by using the representation shown in Figure 3.14. This presentation uses the projections of the representative point P of a composition of the milieu along the sides of the equilateral triangle. Thus, the molar fraction $x_1$ of component $A_1$ will be represented by the length of segment PB in Figure 3.17 (B is the projection of point P along the side $A_2A_3$), and the molar fraction $x_3$ of component $A_3$ will be represented by length PK (K is the projection of point P along the side $A_1A_2$).

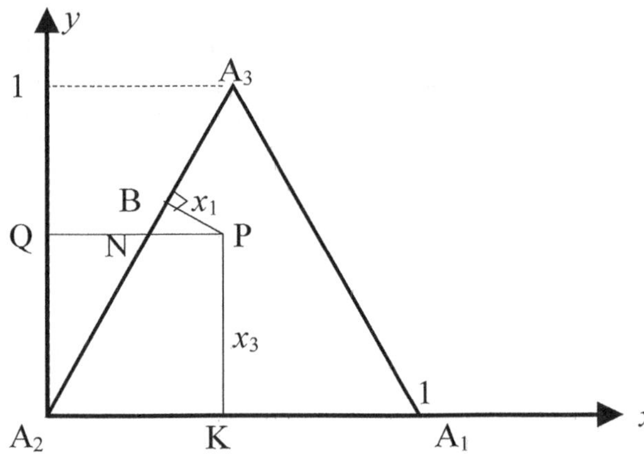


**Figure 3.14.** *Ternary representation of a chemical equilibrium between three gases or three components of a solution*

We shall use a certain number of characteristic curves in that diagram, which we shall define by their equations in the case of perfect solutions. To represent those curves, we shall use the system of rectangular axes defined by an axis $A_2x$ identical to the side $A_2A_1$ and an axis $A_2y$ perpendicular at $A_2$ to the axis $A_2$x (Figure 3.14). The unit borne on each of the axes will be the height of the equilateral triangle.

In order to find the equation of a curve expressed in the system of perpendicular axes $A_2xy$, we first need to be capable of transposing a point P characterized by two molar fractions into coordinates $A_2xy$. Thus, let us consider a point P in the equilateral triangle characterized by the coordinates $x_1$ and $x_3$ (Figure 3.14). We can write:

$$x = PQ = QN + NP \quad [3.70]$$

However, in the triangle $NQA_2$, we have a 30° angle at $A_2$, and therefore:

$$QN = \frac{QA_2}{\sqrt{3}} = \frac{x_3}{\sqrt{3}} \quad [3.71]$$

and in the triangle MPN, we have a 30° angle at P:

$$PB = \frac{\sqrt{3}PN}{2} \quad [3.72]$$

Thus, in view of the value of PB:

$$PN = \frac{2x_1}{\sqrt{3}} \quad [3.73]$$

By feeding back expressions [3.71] and [3.73] into equation [3.70], we find:

$$x = \frac{x_3 + 2x_1}{\sqrt{3}} \quad [3.74]$$

Conversely, we can easily calculate:

$$x_3 = y \quad [3.75a]$$

$$x_1 = \frac{x\sqrt{3} - y}{2} \quad [3.75b]$$

and:

$$x_2 = \frac{2 - y - x\sqrt{3}}{2} \quad [3.75c]$$

For our reaction, we shall define its reaction quotient $Q$ (we use $Q^{(f)}$ or $Q^{(P)}$ obtained on the basis of the fugacities or partial pressures for mixtures of gases) by:

$$Q = \frac{a_3^{\beta_3}}{a_1^{\beta_1} a_2^{\beta_2}} \qquad [3.76]$$

At equilibrium, this reaction coefficient is equal to the equilibrium constant $K_r$:

$$Q^{(\text{equ})} = \frac{a_3^{\beta_3}}{a_1^{\beta_1} a_2^{\beta_2}} = K \qquad [3.77]$$

If the polycomponent phase is a perfect solution, then at equilibrium we have:

$$Q^{(\text{l})} = \frac{x_3^{\beta_3}}{x_1^{\beta_1} x_2^{\beta_2}} = K^{(\text{l})} \qquad [3.78]$$

and if the components are perfect gases, using relation [3.10], then at equilibrium we shall have:

$$Q^{(\text{l})} = \frac{x_3^{\beta_3}}{x_1^{\beta_1} x_2^{\beta_2}} = K^{(\text{l})} = \frac{K^{(P)}}{\left(\frac{P}{P^0}\right)^{\sum \beta_i}} \qquad [3.79]$$

The term $\sum \beta_i$ is defined by the relation:

$$\sum \beta_i = \beta_3 - (\beta_1 + \beta_2) \qquad [3.80]$$

In the triangular diagram, we represent the "iso-parametric" or "iso-Q" curves, which are not curves *per se*, formed of equilibrium points, but some of them, at a given pressure, when relation [3.78] is respected, are the site of points of composition at equilibrium at a certain temperature.

When we start with a certain composition of the reagents at the initial time, with that composition being represented by a point M on the segment

$A_2A_1$, then the system evolves as the reaction progresses, in accordance with very specific compositions set by the initial conditions and the stoichiometric numbers $\beta_1$, $\beta_2$ and $\beta_3$. The curve, passing through a specific point M and giving all the compositions of the system during the reaction, is called the *iso-composition curve*.

At a given pressure and temperature, the iso-composition corresponding to the chosen initial state intersects the iso-Q curve at point E representing the state of thermodynamic equilibrium at the chosen temperature and pressure and for the chosen initial composition.

We shall quantify those data by establishing the iso-Q and iso-composition curves in the case of perfect solutions. First, though, we shall demonstrate an important property of equilibria with perfect solutions.

### 3.5.2. *Molar fractions at equilibrium and initial composition*

We consider equilibria which involve a poly-component perfect phase, and demonstrate the following theorem:

THEOREM 3.1– In such equilibria, the molar fraction of a formed component is, at thermodynamic equilibrium, maximal when that equilibrium is achieved on the basis of the reagents taken in stoichiometric proportions.

This theorem is very general, as long as several components belong to the same phase. We shall now demonstrate it in the case of the three-component reactions given by balance equation [3R.26].

Hence, we consider the initial mixture of a mole and containing $u$ moles of $A_1$ and $1-u$ moles of $A_2$. Let $z$ denote the number of moles of $A_3$ formed. We can see that the quantities of the different components are given by:

$$n_1 = u - z \qquad [3.81]$$

$$n_2 = 1 - u - \frac{\beta_2}{\beta_1} z \qquad [3.82]$$

$$n_3 = \frac{\beta_3}{\beta_1} z \qquad [3.83]$$

and the total quantity of matter is:

$$n = 1 + \frac{\sum \beta_i}{\beta_1} z \qquad [3.84]$$

Thus, the molar fraction $x_3$ of $A_3$ is:

$$x_3 = \frac{z\beta_3 / \beta_1}{1 + \frac{\sum \beta_i}{\beta_1} z} \qquad [3.85]$$

This function is an increasing function of $z$, so the way of finding the maximum of $x_3$ is identical to that for finding the maximum of $z$.

Let us write the law of mass action, feeding back the molar fractions into relation [3.78]:

$$\frac{z^{\beta_3}}{(u-z)^{\beta_1}\left(1-u-\frac{\beta_2}{\beta_1}z\right)^{\beta_2}} = \text{Const.}\left(1+\frac{\sum \beta_i}{\beta_1}z\right)^{\sum \beta_i} \qquad [3.86]$$

By taking the logarithmic derivative of both sides in relation to $u$, and by cancelling all the terms d$z$/d$u$, we obtain:

$$-\frac{\beta_1}{u-z} + \frac{\beta_2}{1-u-\frac{\beta_2}{\beta_1}z} = 0 \qquad [3.87]$$

From this, we deduce that:

$$u = \frac{\beta_1}{\beta_1 + \beta_2} \qquad [3.88a]$$

and:

$$1 - u = \frac{\beta_2}{\beta_1 + \beta_2} \qquad [3.88b]$$

This demonstrates our theorem.

### 3.5.3. *Iso-Q curves in perfect solutions*

To find the general equation for iso-parametric curves, or "iso-Q" curves, we feed the values of $x_1$, $x_2$ and $x_3$ as functions of $x$ and $y$ (expressions [3.75]) back into the application of the law of mass action [3.77]. For the equation of iso-Q curves, we find:

$$\frac{y^{\beta_3}}{\left(x\sqrt{3}-y\right)^{\beta_1}\left(2-y-x\sqrt{3}\right)^{\beta_2}}.2^{(\beta_1-\beta_2)}=Q_x \qquad [3.89]$$

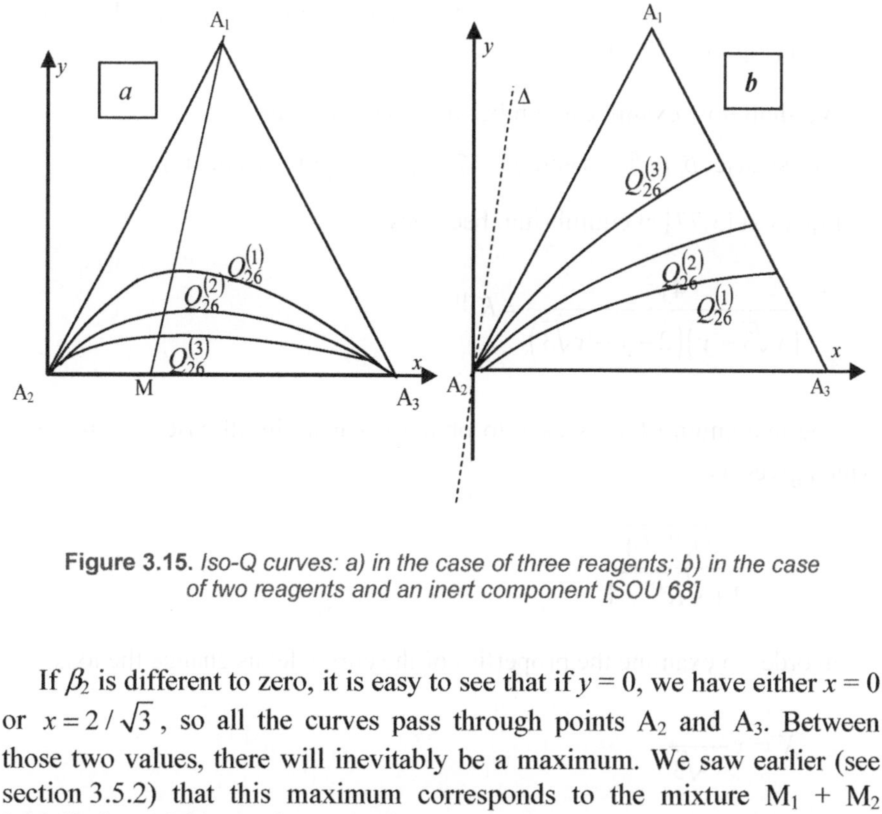


**Figure 3.15.** *Iso-Q curves: a) in the case of three reagents; b) in the case of two reagents and an inert component [SOU 68]*

If $\beta_2$ is different to zero, it is easy to see that if $y = 0$, we have either $x = 0$ or $x = 2/\sqrt{3}$, so all the curves pass through points $A_2$ and $A_3$. Between those two values, there will inevitably be a maximum. We saw earlier (see section 3.5.2) that this maximum corresponds to the mixture $M_1 + M_2$ initially in stoichiometric proportions, meaning that to begin with, point M (Figure 3.15(a)) is such that: $MA_1/MA_2 = \beta_2/\beta_1$. Even after the reaction, this ratio is preserved, so that the space of the maxima of the iso-Q curves will be

$A_3M$ (the ratio of the distances from a point on that line to the sides $A_3A_1$ and $A_3A_2$ is, indeed, constant and equal to $MA_1/MA_2$).

If $\beta_2$ is zero (with component $A_2$ being inert), the iso-Q curves pass through $A_2$. In order to know whether they can have a maximum, we take the logarithmic derivatives by making $\beta_2 = 0$ and d$y$ = 0, which leads us to:

$$-\frac{\beta_1}{x\sqrt{3}-y}=0 \qquad [3.90]$$

This equation is impossible: the curves only exhibit a maximum if the polycomponent phase contains only two components that are involved in the reaction (Figure 3.15(a)).

We shall now examine a number of remarkable cases:

– First case: $\beta_1 = \beta_2 = 1$ and $\beta_3 = 2$ (case of synthesis of IH)

Equation [3.77] at equilibrium becomes:

$$\frac{4y^2}{\left(x\sqrt{3}-y\right)\left(2-y-x\sqrt{3}\right)}=K^{(\mathrm{I})}$$

The maximum of $x_3$ is easy to obtain because, in all cases, $x_m = 1/\sqrt{3}$, which gives us:

$$y_m=\frac{\sqrt{K^{(\mathrm{I})}/4}}{1+\sqrt{K^{(\mathrm{I})}/4}}$$

In order to examine the properties of the curve, let us change the axis:

$$X = x-\frac{1}{\sqrt{3}}$$

In this new coordinate system, the equation of the curve becomes:

$$X^2=\frac{K^{(\mathrm{I})}\left(1-y^2\right)-4y^2}{\left(3K_r\right)}$$

A change from $X$ into $-X$ does not alter $y$. Therefore, the straight line $MA_3$ is the axis of symmetry of the different curves.

– Second case: $\beta_1 = 1$ $\beta_3 = 2$ and $\beta_2 = 0$ (case of reduction of $CO_2$ by carbon in the presence of nitrogen)

Equation [3.77] becomes:

$$\frac{2y^2}{x\sqrt{3}-y} = K^{(\mathrm{I})}$$

From this equation, we draw $x$, which gives us:

$$x = \frac{1}{\sqrt{3}}\left(\frac{2y^2 + K^{(\mathrm{I})}y}{K^{(\mathrm{I})}}\right)$$

The curves are parabolas (Figure 3.15(b)), with their axis parallel to $A_2A_1$, and a peak with a the coordinates ($x = -K^{(\mathrm{I})}/\sqrt{3}$, $y = -2K^{(\mathrm{I})}$). These coordinates are in the ratio $2/\sqrt{3}$. The space of the peaks is therefore the negative part of the line $\Delta$ passing through the origin, with the slope $2/\sqrt{3}$ (Figure 3.15(b)).

– Third case: $\beta_1 = \beta_3 = 4$ and $\beta_2 = 0$ (reduction of $Fe_3O_4$ into metal iron by hydrogen in the presence of nitrogen)

Equation [3.77] becomes:

$$\left(\frac{2y}{x\sqrt{3}-y}\right)^4 = K^{(\mathrm{I})}$$

We deduce the equation for the curves:

$$y = \frac{x\sqrt{3}}{2+K_r^{1/4}}\left(K^{(\mathrm{I})}\right)^{1/4}$$

These curves are straight lines issuing from $A_2$.

The relative positions of the equilibrium curves depend on the sign of the enthalpy of the reaction. If, for instance, the reaction enthalpy is positive (i.e.

if it is an endothermic reaction), then the equilibrium curves will approach point $A_3$ as the temperature increases.

### 3.5.4. *Iso-composition curves in perfect solutions*

Based on the expressions of the quantities of each of the species if we begin with an initial ratio $u$ between $A_1$ and $A_2$ (relations [3.81], [3.82] and [3.83]) we calculate:

$$x_3 = \frac{z\beta_3}{\beta_1 + z\sum \beta_i} = y \qquad [3.91]$$

and:

$$x_1 = \frac{u - z}{1 + \frac{z}{\beta_1}\sum \beta_i} \qquad [3.92]$$

From equation [3.91], we deduce:

$$z = \frac{\beta_1 y}{\beta_3 - y\sum \beta_i} \qquad [3.93]$$

By substituting back into relation [3.92], we deduce the equation of the iso-composition curves:

$$x\sqrt{3} - 2u = y\left(1 - 2\frac{\beta_1 + u\sum \beta_i}{\beta_3}\right) \qquad [3.94]$$

This is the equation of a straight line whose slope depends on the initial conditions (value of $u$).

Consider the intersection of two of those straight lines, obtained for two arbitrary values of $u$: $u_1$ and $u_2$. Applying relation [3.94] for the two values $u_1$ and $u_2$, we find:

$$x\sqrt{3} - 2u_1 = y\left(1 - 2\frac{\beta_1 + u_1\sum \beta_i}{\beta_3}\right) \qquad [3.95]$$

and:

$$x\sqrt{3}-2u_2=y\left(1-2\frac{\beta_1+u_2\sum\beta_i}{\beta_3}\right) \quad [3.96]$$

At the intersection of the two lines, the values of the $x$ and $y$ coordinates are the same, which, if we subtract the two equations [3.95] and [3.96], term by term, leads us to:

$$2(u_2-u_1)=2y_I\frac{\sum\beta_i}{\beta_3}(u_2-u_1) \quad [3.97]$$

This gives us the coordinates of the intersection:

$$y_I=\frac{\beta_3}{\sum\beta_i} \quad [3.98a]$$

and:

$$x_I=\frac{\beta_3-2\beta_1}{\sqrt{3}\sum\beta_i} \quad [3.98b]$$

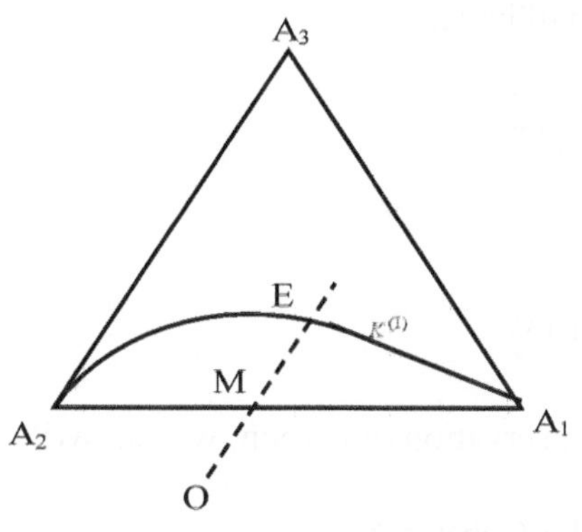


**Figure 3.16.** *Plot of the iso-composition line and the equilibrium point [SOU 68]*

This intersection does not depend on the values of $u_1$ and $u_2$, meaning that the point of intersection O, defined above and whose coordinates are given by equations [3.98a] and [3.98b], is common to all the iso-composition curves. The position of O depends only on the stoichiometric numbers in the balance equation.

The equilibrium point E for the system under examination here (Figure 3.16) is therefore at the intersection of the iso-composition curve – the straight line Om (with m being fixed by the initial mixture $u$) and the iso-$Q$ curve corresponding to the value of the equilibrium constant $K^{(l)}$ at the temperature in question. All of this is calculated for a given pressure, because the diagram is isobaric.

## 3.6. Quaternary diagrams of chemical equilibria

It is possible to plot diagrams for equilibria involving a polycomponent phase with four components of the reaction. The system contains three independent components. These diagrams are square. To demonstrate their representation and their operation, we shall consider the reduction of carbon dioxide by hydrogen, in accordance with:

$$CO_2 + H_2 = CO + H_2O \qquad [3R.27]$$

On the horizontal abscissa axis, we shall show the proportion of carbon in the CO state, and on the vertical axis, the proportion of $H_2$ in the state of free $H_2$ gas. Hence, we shall have:

$$x = \frac{\{CO\}}{\{CO\} + \{CO_2\}} \qquad [3.99a]$$

$$y = \frac{\{H_2\}}{\{H_2\} + \{H_2O\}} \qquad [3.99b]$$

Because of the conservation of carbon, we can write:

$$\{CO\} + \{CO_2\} = \text{Const.} = k_C \qquad [3.100]$$

Similarly, for the conservation of hydrogen:

$$\{H_2\} + \{H_2O\} = \text{Const.} = k_H \qquad [3.101]$$

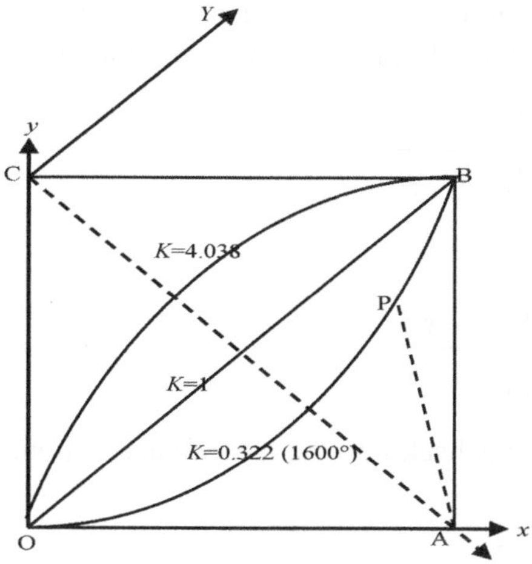


**Figure 3.17.** *Square diagram for chemical equilibrium between four gases [SOU 68]*

Thus, we deduce from it the expressions:

$$1-x=\frac{\{CO_2\}}{\{CO_2\}+\{CO\}} \quad [3.102a]$$

and:

$$1-y=\frac{\{H_2O\}}{\{H_2\}+\{H_2O\}} \quad [3.102b]$$

Figure 3.17 gives the square diagram of side length of one unit, and each peak of which represents a mixture of two gases as shown in Table 3.1, with the proportions being variables.

| O | A | B | C |
|---|---|---|---|
| Mixture $CO_2 + H_2O$ | $CO + H_2O$ | $CO + H_2$ | $CO_2 + H_2$ |

**Table 3.1.** *Meaning of each of the corners of the square diagram*

We shall apply the law of mass action expressed in relation to the partial pressures at equilibrium. In order to do so, we note that:

$$\frac{P_{H_2}}{P_{H_2O}} = \frac{\{H_2\}}{\{H_2O\}} = \frac{y}{1-y} \qquad [3.103]$$

and that:

$$\frac{P_{CO}}{P_{CO_2}} = \frac{\{CO\}}{\{CO_2\}} = \frac{x}{1-x} \qquad [3.104]$$

By substituting this back into the law of mass action, we find:

$$\frac{(1-x)y}{(1-y)x} = K_{27}^{(P)} \qquad [3.105]$$

This gives us the general equation for the curves, which are sets of points of equilibrium for reaction [3R.27]:

$$y = \frac{K_{27}^{(P)} x}{\left(K_{27}^{(P)}+1\right)x+1} \qquad [3.106]$$

These curves (see Figure 3.17) are equilateral hyperbolas with a horizontal asymptote and a vertical asymptote. The concavity of the branch within the square is downturned if the constant $K_{27}^{(P)}$ is greater than 1 and upturned if that constant is less than 1. If $K_{27}^{(P)}$ is equal to 1, then the hyperbola is degenerated into a straight line, which is the main diagonal of the square.

These curves can have an axis of symmetry which is the second diagonal of the square (Figure 3.17) as shown by the change of variables in accordance with:

$$Y = \frac{x-(1-y)}{\sqrt{2}} \qquad [3.107a]$$

and:

$$X = \frac{x+(1-y)}{\sqrt{2}} \qquad [3.107b]$$

Hence, we are led to the expression of the old coordinates as a function of the new ones:

$$x = \frac{X + Y}{\sqrt{2}} \quad [3.108a]$$

and:

$$y = 1 + \frac{Y - X}{\sqrt{2}} \quad [3.108b]$$

This gives us the new equation for the curves:

$$X^2\left(1 - K_{27}^{(P)}\right) - 2\sqrt{2}X = Y^2\left(1 - K_{27}^{(P)}\right) - 2 \quad [3.109]$$

If, in that equation, we change $Y$ into $-Y$, $X$ does not change, which proves the symmetry in relation to the second diagonal.

Let us demonstrate the following property: in every transformation (isothermal or otherwise), the space of points representative of those intermediary states is a straight line with a slope of $-k_C/k_H$.

Indeed, when we take account of relations [3.99], [3.100] and [3.101], we have:

$$x = \frac{\{CO\}}{k_C} \quad [3.110a]$$

and:

$$y = \frac{\{H_2\}}{k_C} \quad [3.110b]$$

During the course of an elementary evolution of the extent of the reaction, the coordinates of the figurative point will vary by $dx$ and $dy$, such that:

$$dx = \frac{d\{CO\}}{k_C} \quad [3.111a]$$

and:

$$dy = \frac{d\{H_2\}}{k_C} \qquad [3.111b]$$

However, because of the stoichiometry of the reaction, the elementary variations of the quantities of matter are not random, and necessarily we have:

$$d\{CO\} = -d\{H_2\} \qquad [3.112]$$

and therefore, by substituting back into relations [3.111], we obtain:

$$\frac{dy}{dx} = \frac{k_C}{k_H} \qquad [3.113]$$

This is the equation of a straight line. The line passes through the point representative of the initial mixture.

To demonstrate the use of the diagram, let us begin with a mixture formed of 5 moles of carbon monoxide and 1 mole of water, and determine the equilibrium compositions at 1600°C – a temperature where the constant $K_{27}^{(P)}$ is 0.322. The initial point of the mixture is at A according to Table 3.1. The slope of the line representing the evolution is –5. The equilibrium at 1600°C is represented by point P, which is the point of intersection of the straight line starting at A with the slope of –5 and the equilibrium isotherm with the constant 0.322 (Figure 3.17). The coordinates of P are $x = 0.87$ and $y = 0.62$. This gives us the composition at equilibrium:

– 5 × 0.87 = 4.35 moles of CO and therefore 0.65 moles of $CO_2$;

– 1 × 0.62 = 0.62 moles of $H_2$ and therefore 0.38 moles of $H_2O$.

In this chapter, we have defined and used a number of equilibrium constants. The next chapter will be devoted to the evaluation of the numerical values of those constants, either by experimentation or by computation.

4

# Determination of the Values Associated with Reactions – Equilibrium Calculations

The objective of this chapter is, by experimentation or computation, on the basis of the values of the main properties, to determine the value of the equilibrium constant of a reaction, or, which is the same thing by virtue of relation [4.2], that of the standard molar Gibbs energy associated with the reaction in question.

We shall then examine the use of those values to calculate complex equilibria, enabling us to define domains of stability of the phases and the compounds, domains of predominance of components, and to serve the ultimate goal of a transformation within the system.

In this chapter, we only look at molecular reactions. Ionic equilibria will be studied in Volume 6 of this Set of books.

## 4.1. Reminders of a few thermodynamic relations

In this section, we shall establish a certain number of thermodynamic relations specific to the reactions, which will be useful to us in this chapter, particularly, for calculating the values.

Remember that every time we attempt to determine an equilibrium constant, we need to begin by choosing the constant in question (see section 3.1.2) and the reference states. Remember that as a reference, we can choose either the standard state, in which the different compounds are pure at the pressure of 1 bar, in their normal state of condensation at the chosen temperature (reference I) or a reference state in which the solutions are infinitely dilute at the pressure of 1 bar (reference II), or indeed, for solutions, a state in which all the solutes are at the concentration of 1 mol/l (reference III).

The Gibbs energy in the reference state associated with a reaction is linked to its enthalpy and its entropy in the same reference state at temperature $T$ *by*:

$$\Delta_r G_T^0 = \Delta_r H_T^0 - T\Delta_r S_T^0 \quad [4.1]$$

The equilibrium constant of a reaction in the chosen reference state is defined on the basis of its corresponding associated Gibbs energy by:

$$K = \exp - \frac{\Delta_r G_T^0}{\mathrm{R}T} \quad [4.2]$$

This equilibrium constant varies with temperature, in accordance with the van 't Hoff equation:

$$\frac{\partial \ln K}{\partial T} = \frac{\Delta_r H_T^0}{\mathrm{R}T^2} \quad [4.3]$$

The enthalpy of the reaction $r$ varies with temperature, according to the Kirschoff relation [4.4] as a function of the specific heat capacity at constant pressure:

$$\left(\frac{\partial \Delta_r H^0}{\partial T}\right)_P = \Delta_r C_P^0 \quad [4.4]$$

The term $\Delta_r C_P^0$ is linked to the molar specific heat capacities at constant pressure in the reference conditions of all the components of the reaction by the expression [4.6]:

$$\Delta_r C_P^0 = \sum_k \nu_k C_{P(k)}^0 \quad [4.5]$$

The entropy of the reaction varies with temperature in accordance with relation [4.6], as a function of the specific heat capacity at constant pressure and of the temperature:

$$\left(\frac{\partial \Delta_r S}{\partial T}\right)_P = \frac{\Delta_r C_P}{T} \qquad [4.6]$$

If we consider a substance undergoing an allotropic transformation in the solid state at temperature $T_a$, which melts at temperature $T_F$ and boils at temperature $T_{Eb}$, to integrate expressions [4.4] and [4.6], it is necessary to divide the temperature interval between the initial temperature $T_0$ and the temperature $T$ into slices. Each slice is characterized by a phase and therefore a function of the molar specific heat capacity at constant pressure with changing temperature. Thus, integration of equation [4.4] involves two types of terms:

1) integral terms in the form $\int_{T_1}^{T_2} \Delta C_P^{(\varphi)}\, \mathrm{d}T$ in the domain of stability of phase $\varphi$;

2) terms due to the changes which stem from the enthalpies associated with the phase changes of the components in the reaction. For the phase change $\Delta\varphi$ of component $A_k$, whose algebraic stoichiometric number in the reaction is $\nu_k$, the corresponding term is: $\nu_k \Delta_{\Delta\varphi} H_k$.

Thus, the enthalpy takes the form of a sum similar to equation [4.7], in which the number of terms depends on the initial temperature $T_0$ at which we know the enthalpy, and on the number of state changes of the substance between the temperature $T_0$ and the effective temperature $T$.

$$\Delta_r H_T = \Delta_r H_0 + \sum_{\varphi} \int_{T_{\Delta\varphi}}^{T} \Delta_r C_P^{(\varphi)}\, \mathrm{d}T + \sum_{\Delta\varphi, k} \nu_k \Delta_{\Delta\varphi} H_k \qquad [4.7]$$

We proceed in the same way for the integration of the entropy function on the basis of relation [4.6], with the same division of the temperature range. We find integral terms in the form $\int_{T_1}^{T_2} \frac{\Delta_r C_P^{(\varphi)}}{T} \mathrm{d}T$ and terms of state change in the form: $\nu_k \dfrac{\Delta_{\Delta\varphi} H_k}{T^{(\Delta\varphi)}}$. Thus, we obtain sums of terms such as those illustrated by:

$$\Delta_r H_T = \Delta_r H_0 + \sum_{\varphi} \int_{T_{\Delta\varphi}}^{T} \frac{\Delta_r C_P^{(\varphi)}}{T} \mathrm{d}T + \sum_{\Delta\varphi,k} \nu_k \frac{\Delta_{\Delta\varphi} H_k}{T} \qquad [4.8]$$

Expressions [4.7] and [4.8] need to be applied to each individual case.

Next, we can combine expressions [4.7], [4.8], [4.1] and [4.2] and then integrate relation [4.3] to obtain an expression of the rate constant at temperature $T$, which is in the form:

$$\begin{aligned}\ln K_{P(T)} = &-\frac{\Delta_r H_{T_0}^0}{RT} - \frac{1}{\mathrm{RT}} \sum_{\varphi} \int_{T_{(\Delta\varphi)}}^{T} \Delta_r C_P^{(\varphi)} \mathrm{d}T - \frac{1}{RT} \sum_{\Delta\varphi,k} \nu_k \Delta_{\Delta\varphi} H_k \\ &+ \frac{\Delta_r S_{T_0}^0}{\mathrm{R}} + \frac{T}{\mathrm{R}} \sum_{\varphi} \int_{T_{(\Delta\varphi)}}^{T} \frac{\Delta_r C_P^{(\varphi)}}{T} \mathrm{d}T + \frac{T}{\mathrm{R}} \sum_{\Delta\varphi,k} \nu_k \frac{\Delta_{\Delta\varphi} H_k}{T^{(\Delta\varphi)}}\end{aligned} \qquad [4.9]$$

By combining all these relations, we can also obtain an expression which links the equilibrium constant at a temperature $T$ to a value of that constant at a different $T_0$, and the variations of the molar specific heat capacities in each temperature interval defined. We finally find a relation of the type:

$$\begin{aligned}\mathrm{R}T \ln K_{P(T)} = &\mathrm{R}T_0 \ln K_{P(T_0)} - \sum_{\varphi} \int_{T_{(\Delta\varphi)}}^{T} \Delta_r C_P^{(\varphi)} \mathrm{d}T - \sum_{\Delta\varphi,k} \nu_k \Delta_{\Delta\varphi} H_k \\ &+ (T - T_0) \Delta_r S_{T_0}^0 + T \sum_{\varphi} \int_{T_{(\Delta\varphi)}}^{T} \frac{\Delta_r C_P^{(\varphi)}}{T} \mathrm{d}T + T \sum_{\Delta\varphi,k} \nu_k \frac{\Delta_{\Delta\varphi} H_k}{T^{(\Delta\varphi)}}\end{aligned} \qquad [4.10]$$

To illustrate this approach, we shall determine an expression of the vapor pressure of liquid copper at temperature $T$, which is above its fusion point of 1357 K. As a reference state, we choose the standard state of the pure substance at a temperature of 298 K and pressure of 1 bar. We have the following data:

– standard enthalpy of vaporization of liquid copper at 298 K: $\Delta_V h_{298}^0 = 19498$ J.mole$^{-1}$;

– standard entropy of vaporization of liquid copper at 298 K: $\Delta_V s_{298}^0 = 7.60 \text{ J.mol}^{-1} K^{-1}$;

– enthalpy of fusion of copper at 1357 K: $\Delta_F h_{1357} = 741.6 \text{ J.mol}^{-1}$

– specific heat capacity of solid copper at constant pressure:

$$C_P^{(\text{sol})} = 1.30 + 3.52.10^{-3} T \text{ J.mol}^{-1}\text{K}^{-1}$$

– specific heat capacity of liquid copper at constant pressure, independent of the temperature:

$$C_P^{(\text{liq})} = 1.79 \text{ J.mol}^{-1}\text{K}^{-1}$$

– specific heat capacity of copper vapor at constant pressure, independent of the temperature:

$$C_P^{(vap)} = 1.189 \text{ J.mol}^{-1}\text{K}^{-1}$$

We shall express the term $\Delta_{(\text{sublim})} C_P = C_P^{(\text{vap})} - C_P^{(\text{sol})}$ between 298 and 1357 K:

$$\Delta_{(\text{sublim})} C_P = -0.111 - 3.52.10^{-3} T$$

Let us express the term $\Delta_\text{v} C_P = C_P^{(vap)} - C_P^{(liq)}$ between 1357 K and temperature $T$:

$$\Delta_\text{v} C_P = -0.605 \text{ J.mol}^1 K^{-1}$$

The standard enthalpy at $T$ will be given by:

$$\Delta_\text{v} h_T^0 = \Delta_\text{v} h_{298}^0 + \int_{298}^{1357} C_P^{(sol)} \,\text{d}T + \Delta_\text{F} h_{1357} + \int_{1357}^{T} \Delta_\text{v} C_P \,\text{d}T$$

Thus, when we feed the specific heat capacities and the enthalpy of fusion back into this formula, we obtain:

$$\Delta_\text{v} h_T^0 = \Delta_\text{v} h_{298}^0 + \int_{298}^{1357} \left(-0.111 - 3.52.10^{-3} T\right) \text{d}T + 741.6 - \int_{1357}^{T} 0.605 \,\text{d}T$$

This gives us the expression:

$$\Delta_{\mathrm{v}} h_T^0 = 11858 - 0.065T$$

Similarly, the standard entropy at $T$ will be given by:

$$\Delta_{\mathrm{v}} s_T^0 = \Delta_{\mathrm{v}} s_{298}^0 + \int_{298}^{1357} \frac{C_P^{(sol)}}{T} \mathrm{d}T + \Delta_{\mathrm{F}} s_{1357} + \int_{1357}^{T} \frac{\Delta_{\mathrm{v}} C}{T}_P \mathrm{d}T$$

By feeding the values of the specific heat capacities into this equation, we find:

$$\Delta_{\mathrm{v}} s_T^0 = \Delta_{\mathrm{v}} s_{298}^0 + \int_{298}^{1357} \frac{\left(-0.111 - 3.52.10^{-3} T\right)}{T} \mathrm{d}T + \frac{741.6}{1357} - \int_{1357}^{T} \frac{0.605}{T} \mathrm{d}T$$

This expression can be simplified to:

$$\Delta_{\mathrm{v}} s_T^0 = 761 - 0.605 \ln T$$

Ultimately, the standard Gibbs energy is:

$$\Delta_{\mathrm{v}} g_T^0 = \Delta_{\mathrm{v}} h_T^0 - T \Delta_{\mathrm{v}} s_T^0 = 11858 - 760.9T + 0.605T \ln T$$

This gives us a value for the saturating vapor pressure, expressed in bars, which is the equilibrium constant of the vaporization reaction:

$$\ln P_{\mathrm{Cu}}^{(\mathrm{vap})} = \ln K^{(1)} = -\frac{\Delta_{\mathrm{v}} g_T^0}{\mathrm{R}T} = -91.45 + 0.072 \ln T - \frac{1425}{T}$$

As we can see, the expressions with which we are working here can quickly become rather complex, which is why it is so helpful to use computer technology to perform the calculations.

## 4.2. Enthalpies of reaction – thermochemistry

We know that the reaction enthalpy is identical to the reaction heat at constant pressure. The properties and means of determining these reaction heats make up *hermochemistry*.

### 4.2.1. *Experimental determination of the reaction enthalpies by calorimetry*

The development of knowledge of reaction enthalpies is closely linked to that of reaction heats, and therefore to calorimetry and thermochemistry.

Experimental measurement of a reaction heat at constant pressure at a temperature determined by calorimetry requires the measurement to respect a set of essential precautions:

– the first condition is that the tester must control the triggering of the reaction so that it does not start before the measurements begin to be taken;

– the reaction must be sufficiently quick as to enable us to measure the heat involved over the course of the whole reaction without the losses of heat from the calorimeter due to radiation interfering with the measurement;

– the reaction must not be accompanied by secondary reactions for which the reactions' heats are not known; and

– the products used must be sufficiently pure.

These conditions mean that direct measurement of the heat of a large number of reactions is still tricky, and therefore a significant amount of data still needs to be improved.

### 4.2.2. *Calculation of the standard enthalpy at another temperature*

If we know the enthalpy at a temperature $T_0$, it is easy to use a relation such as [4.7] to calculate that enthalpy at any temperature $T$, provided we know the variations of the molar specific heat capacities of the reaction components between temperatures $T_0$ and $T$ and the enthalpies of state change of those components within that temperature range.

### 4.2.3. *Influence of the pressure on the reaction enthalpies*

Consider the general relation giving the variations of enthalpy with pressure:

$$\frac{\partial H}{\partial P} = PV \qquad [4.11]$$

From this relation, we deduce the enthalpy attached to a reaction *r*:

$$\frac{\partial \Delta_r H}{\partial P} = P\Delta_r V \qquad [4.12]$$

By integrating this relation [4.12], we are able to calculate the value of an enthalpy, associated with a reaction, at a pressure *P*, when we know its value at another pressure and the relation between the volume and pressure (e.g. the equation of state) for each of the components of the reaction.

Note, though, that:

– for reactions involving condensed substances, their contribution to the term $\Delta_r V$ is negligible within the usual pressure range; and

– as for gases, provided they are considered perfect, the term $P\Delta_r V$ is $(\Sigma\nu)RT$ at constant temperature.

### 4.2.4. *Determination of the reaction enthalpies by calculation on the basis of other thermodynamic data*

There are three main methods used to calculate an enthalpy associated with a reaction on the basis of other thermodynamic data. These methods entail using:

– the variation of the Gibbs energy associated with the reaction with temperature;

– the variation of the equilibrium constant of that reaction with temperature; or

– indeed appropriately-chosen enthalpy values.

#### 4.2.4.1. *Calculation of an enthalpy on the basis of a value of Gibbs energy*

This method is based on Helmholtz's law, which specifies the variations of Gibbs energy with temperature:

$$\frac{\mathrm{d}\left(\frac{G}{T}\right)}{\mathrm{d}T} = -\frac{H}{T^2} \qquad [4.13]$$

We suppose variations of the standard Gibbs energy of the reaction as a function of the temperature to be known. Based on relation [4.13], we deduce the standard enthalpy associate, using the relation:

$$\Delta_r h^0 = -T^2 \frac{\mathrm{d}\left(\dfrac{\Delta_r g^0}{T}\right)}{\mathrm{d}T} \qquad [4.14]$$

If we consider two temperatures $T_1$ and $T_2$ which are not too far apart, then knowledge of the standard Gibbs energy at those two temperatures enables us to approximately calculate the standard enthalpy at another temperature lying between $T_1$ and $T_2$, by the relation:

$$\Delta_r h^0 = \frac{T_1 T_2}{T_2 - T_1}\left(\frac{\Delta_r g_1^0}{T_1} - \frac{\Delta_r g_2^0}{T_2}\right) \qquad [4.15]$$

### 4.2.4.2. *Calculation of an enthalpy on the basis of the equilibrium constant*

This method is based on the van 't Hoff equation for the variation of the equilibrium constants with temperature, which we shall suppose to be known. From relation [4.3], we deduce the associated enthalpy:

$$\Delta_r h^0 = -\mathrm{R} \frac{d \ln K^{(\mathrm{I})}}{d\left(\dfrac{1}{T}\right)} \qquad [4.16]$$

If we choose two temperatures $T_1$ and $T_2$ which are not too far apart, then we only need to know the equilibrium constant at the two temperatures $T_1$ and $T_2$. Then, the standard enthalpy at a temperature between $T_1$ and $T_2$ is given approximately by:

$$\Delta_r h^0 = \mathrm{R} \frac{T_1 T_2}{T_2 - T_1} \ln \frac{K_2^{(\mathrm{I})}}{K_2^{(\mathrm{I})}} \qquad [4.17]$$

#### 4.2.4.3. *Calculation of an enthalpy on the basis of a set of reaction heats appropriately chosen and known*

This method is based on the first law of thermodynamics, which enables us to deduce that the enthalpy is a function of state. If we choose a certain number of reactions such that – a linear combination of their balance equations gives us the balance equation for the reaction under study; and we know the enthalpies of the different reactions chosen then we know that, between the enthalpy of the sought reaction $\Delta_r H$ and those of the chosen reactions $\Delta_m H$, there is the same linear combination as that which exists between the balance equations in the form:

$$\Delta_r H = \sum_k \lambda_m \Delta_m H \qquad [4.18]$$

This method enables us to determine the enthalpy at a specific temperature provided the enthalpies of the combined reactions are known at that same temperature.

In fact, this method is Hess's long-established method for determining reaction heats.

### 4.2.5. *Enthalpies of formation*

One category of reaction enthalpies plays a very particular role in the methods for determining the standard enthalpy associated with a reaction: the enthalpies of formation.

#### 4.2.5.1. *Definition of the enthalpies of formation*

The *enthalpy of formation* of substance is the enthalpy associated with the reaction to synthesize a mole of that compound from its elements in molecular form.

For example, the enthalpy of formation of methane is the enthalpy associated with the reaction:

$$\text{C(graphite)} + 2\ H_2 = CH_4 \qquad [4R.1]$$

A particular class of enthalpies of formation is constituted by standard enthalpies of formation at a temperature $T$ (written as $H_T^0$), which are those obtained when the compound and its elements are all in their standard state at that temperature. Thus, we have the relation between standard enthalpy of formation and enthalpy associated with the reaction of formation $f$:

$$H_T^0 = \Delta_f h_T^0 \qquad [4.19]$$

We shall see later on that the tables of thermodynamic data give all the values of the standard enthalpies of formation at the temperature of 25°C or 298.15 K: $H_{298}^0$.

#### 4.2.5.2. *Use of the enthalpies of formation to calculate the standard enthalpy of a reaction*

If we know the enthalpies of formation of all the components of a reaction, it is easy, by application of Hess's law (see section 4.2.4.3) to calculate the standard enthalpy of the reaction at the same temperature. Application of relation [4.19] gives us:

$$\Delta_r h_T^0 = \sum_k \nu_k H_{T(k)}^0 \qquad [4.20]$$

NOTE 4.1.– standard enthalpies of formation constitute a set of values of the enthalpies of substances; it is the set of enthalpies at standard pressure for which the origin zero is posited for the enthalpy of simple substances.

#### 4.2.5.3. *Determination of the enthalpies of formation*

We can use three methods to determine the standard enthalpy of formation of a compound at a given temperature $T$:

– The first method is to use calorimetry to experimentally measure the direct synthesis reaction and then to use relation [4.19]. This method, which is apparently simple, is in fact often difficult or even impossible to apply. For example, in the case of methane synthesis, reaction [4R.1] is impossible to perform directly.

– The second family of methods is the use of Hess's theorem, as we saw in section 4.2.4.3. In order to do this, we need to have reactions which will give us the formation reaction when combined linearly, and their standard enthalpies. We shall see later on that the reactions and total heats of combustion are very practical for that use – particularly in organic chemistry, where syntheses from simple substances are rarely possible.

– The third method is to apply relation [4.20] to a reaction involving the compound under study, and for which we know the standard enthalpy of reaction and the enthalpies of formation of the other components involved in the reaction.

NOTE 4.2.– By combining relations [4.19] and [4.20], we can see that, by definition, the standard enthalpies of formation of simple substances such as C, $O_2$, $H_2$, $Cl_2$, Fe, Na, etc., are null at all temperatures.

#### 4.2.5.4. *Enthalpies of formation and temperature*

The fact that the enthalpies of formation of the simple substances are chosen as equal to zero at all temperatures means that the enthalpy of formation will obey relation [4.7] with temperature. This is consistent with the fact that the enthalpies of formation represent a set of values of the enthalpies of the pure substances.

### 4.2.6. *Enthalpies of combustion*

As we noted earlier, it is not always possible to synthesize a pure substance directly, with no secondary reactions, from its elementary substances, and it is then not possible to determine an enthalpy of formation by the direct method. This obstacle is often circumvented by considering the enthalpies of combustion of the species in the presence of excess oxygen.

#### 4.2.6.1. *Definitions*

The *enthalpy of combustion* of a pure substance is the enthalpy associated with the reaction of complete combustion of a mole of that pure substance in the presence of excess oxygen.

The standard enthalpy of combustion at a temperature $T$ is the enthalpy of combustion at the standard pressure of 1 bar and temperature $T$.

#### 4.2.6.2. *Experimental determination of the standard enthalpies of combustion*

The standard enthalpy of combustion is determined on the basis of the heat of combustion measured by calorimetry at standard pressure.

Let $Q^0$ represent the measured heat of combustion of $n_0$ moles of the compound in question at normal pressure, when the reaction starts at temperature $T$ and finishes at the final temperature $T_f$. If $n_{O_2}^{xs}$ moles of the excess oxygen were consumed, and the standard enthalpy of the reaction at temperature $T$ is the measured heat of reaction, expressed in relation to 1 mole, decreased by the amounts of heat which have been used to raise the combustion products and excess oxygen from temperature $T$ to temperature $T_f$:

$$\Delta_{comb} H_T^0 = \frac{Q^0}{n_0} - \sum_{\text{products}} \int_T^{T_f} \nu_{\text{products}} C_{P(\text{products})} \, \mathrm{d}T - \frac{n_{O_2}^{xs}}{n_0} \int_T^{T_f} \nu_{O_2} C_{P(O_2)} \, \mathrm{d}T \qquad [4.21]$$

Of course, if, in the experiment, the final pressure is different from the initial pressure, a pressure correction can potentially be applied – particularly for gaseous species.

#### 4.2.6.3. *Use of the standard enthalpies of combustion to calculate the standard enthalpy of a reaction*

If we consider the balance equation of any reaction $r$, we can always consider it as the linear combination of the reactions of combustion of the different components in the reaction under study, labeled with their stoichiometric coefficients in the reaction in question. We can therefore write the following for the standard enthalpy of reaction:

$$\Delta_r h_T^0 = \sum_k \nu_k \Delta_{comb} H_{T(k)}^0 \qquad [4.22]$$

#### 4.2.6.4. *Use of the standard enthalpies of combustion to calculate the standard enthalpies of formation*

The above calculation can be applied to synthesis reactions, and therefore the enthalpy of formation can be deduced from the combustion reactions, by:

$$H_T^0 = \sum_k \nu_k \Delta_{comb} H_{T(k)}^0 \qquad [4.23]$$

with the sum being extended to the components of the synthesis reaction.

| Substance | Enthalpy of combustion | Substance | Enthalpy of combustion |
|---|---|---|---|
| Methane | 889.46±9.20 | Pentane n | 3533.23±0.88 |
| Ethane | 1570.89±0.46 | Ethylene | 1409.83±0.29 |
| Propane | 2217.78±0.50 | Propylene | 2055.81±0.63 |

**Table 4.1.** *Enthalpy values of combustion in kJ/mol at 25°C. Data from the National Bureau of Standards*

Table 4.1 shows a number of values of the enthalpies of combustion at 25°C, expressed in kilojoules per mole.

The values of the enthalpies of combustion are sometimes tricky to determine, as is the case, for instance, with carbon, because we need to begin with a perfectly pure form of carbon (graphite and diamond). However, note that a high degree of accuracy is necessary, because, to clarify our ideas, a 1 kJ/mol error on an enthalpy of combustion at 25°C leads to an error 10 times greater on an equilibrium constant at a temperature of 656 K.

### 4.2.6.5. *Adiabatic flame temperatures*

The *adiabatic flame temperature* of a compound denotes the *calculated* temperature reached by the reaction products when all of the heat released during the combustion of that compound at a determinate temperature is used to heat those products. The energy balance immediately gives us the relation:

$$\Delta_{\text{comb}} H_T^0 = \int_{T_0}^{T} \nu_k C_{P(\text{product})} \mathrm{d}T \qquad [4.24]$$

We can show that the flame temperature increases if the starting temperature rises. On the other hand, it decreases if the oxygen necessary for combustion is replaced by the equivalent amount of air, because it then becomes necessary to heat the inert nitrogen as well.

For indicative purposes, the adiabatic flame temperature of a mole of carbon monoxide taken at 25°C in just the necessary amount of pure oxygen is in the order of 5058 K. If the oxygen is replaced by air, the adiabatic flame temperature of a mole of carbon monoxide is now only 2663 K.

The true flame temperatures are much lower than the values calculated in this way, because combustion is far from adiabatic.

## 4.2.7. *Dissociation energy, bond energy and enthalpies of formation*

A second way of determining the enthalpies of formation is to use the bond energies.

### 4.2.7.1. *Enthalpy of formation from gaseous monatomic elements*

Up until now, we have considered the energies of formation from the simple substances taken in their normal state in standard conditions. We shall now turn our attention to the enthalpies of formation of the compounds from their constituent atoms in the gaseous state, which we call the *enthalpy of synthesis*. We also find the opposite of that value: the *enthalpy of atomization*, which is the enthalpy released during the complete separation of a substance into single atoms. In order to switch from our standard enthalpies of formation to these enthalpies of synthesis, we need to know the enthalpies of dissociation of the simple substances and the enthalpies of sublimation and vaporization for the solid and liquid monatomic compounds.

### 4.2.7.2. *Energy of sublimation*

Thanks to experimental determination, we know the enthalpies of sublimation of numerous solids at ordinary temperature, such as sulfur, selenium, phosphorus, tin, etc. The only truly tricky point, in fact, is determining the enthalpy of sublimation of carbon in graphite (or diamond) form, about which there have been a great many studies conducted. We cannot be certain of the values that are currently accepted.

### 4.2.7.3. *Energy of dissociation*

Consider the water molecule in the gaseous state. It comprises two OH bonds.

Let us consider the breaking of the first bond by the following reaction:

$$H_2O(gas) = H(gas) + OH(gas) \quad [4R.2]$$

The standard enthalpy of such a reaction is:

$$\Delta_{R2}h^0_{298} = 501.87 \text{ kJ.mol}^{-1}$$

We shall now consider the breaking of the second OH bond – i.e. the following reaction:

$$\text{OH(gas)} = \text{H(gas)} + \text{O(gas)} \quad [4R.3]$$

Its standard enthalpy is:

$$\Delta_{R3}h^0_{298} = 423.38 \text{ kJ.mol}^{-1}$$

Therefore, we can determine the enthalpy of atomization of water by finding the sum of the previous two values.

#### 4.2.7.4. *Determination of enthalpies of atomization*

It is possible to determine the enthalpies of atomization (or dissociation) of simple gaseous substances such as $H_2$, $O_2$, $N_2$, $Cl_2$ $Br_2$, $I_2$, etc. either by examining equilibria in which those substances are involved, or on the basis of spectroscopic data.

We shall look in detail at this latter method, which yields interesting results because it does not require any temperature measurement to be taken.

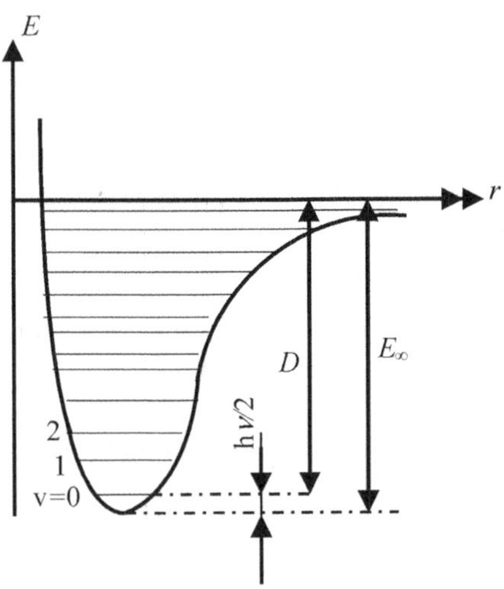


**Figure 4.1.** *Energy of a bond as a function of the inter-atomic distance*

For simplicity's sake, we shall base our discussion on a diatomic molecule. Consider the curve (Figure 4.1) showing the variations of the vibration energy as a function of the distance between the nuclei. At equilibrium between the forces of attraction and repulsion, the nuclei are a distance $r_0$ apart (which is shown by the minimum of the curve).

If the molecule receives energy, a vibratory motion occurs around the equilibrium state, so that the distance between the nuclei oscillates between a maximal value $r_{max}$ and a minimal value $r_{min}$. The vibration is not entirely harmonic, because the forces increase more quickly as the nuclei draw near to one another than they decrease as the nuclei move apart. The vibrational energy which would yield a parabola in the case of a harmonic vibration, in reality, exhibits the asymmetrical form shown in Figure 4.1. The vibrational energy can then be expressed in the form:

$$E_{\mathrm{v}} = (\mathrm{v}+1/2)\mathrm{h}\nu_{\mathrm{vibr}} + (\mathrm{v}+1/2)^{1/2} x\mathrm{h}\nu_{\mathrm{vibr}} \qquad [4.25]$$

This expression includes two terms. The first is due to a harmonic oscillation, and the second is a correction of the first term, and contains the coefficient of anharmonicity $x$ given by:

$$x = \frac{\mathrm{h}\nu_{\mathrm{vibr}}}{4E_{\infty}} \qquad [4.26]$$

The term v, which is an integer that may have a variety of positive values, gives us quantified energies represented by the horizontal lines in Figure 4.1.

The energy difference between two consecutive vibrational terms (i.e. the shift from v to v + 1) is therefore:

$$\Delta_{\mathrm{v}}^{\mathrm{v}+1}E = \mathrm{h}\nu_{\mathrm{vibr}}\left[1-2x(\mathrm{v}+1)\right] = \mathrm{h}\nu_{\mathrm{vibr}}\left[1-2\frac{\mathrm{h}\nu_{\mathrm{vibr}}}{4E_{\infty}}(\mathrm{v}+1)\right] \qquad [4.27]$$

This energy difference has a corresponding absorption band of frequency $\nu_{\mathrm{abs}}$ such that:

$$E = \mathrm{h}\nu_{\mathrm{abs}} \qquad [4.28]$$

The distance (in terms of frequency) between two spectral lines in the same series, obtained for two consecutive values v and v + 1, will therefore be:

$$\Delta_{\mathrm{v}}^{\mathrm{v+1}}\nu_{\mathrm{abs}} = \nu_{\mathrm{vibr}}\left[1-2\frac{\mathrm{h}\nu_{\mathrm{vibr}}}{4E_{\infty}}(\mathrm{v}+1)\right] \qquad [4.29]$$

This distance shrinks as v increases, until it finally disappears entirely. Thus, the vibrational energy cannot surpass a certain limit. If we give the molecule more energy than that limit, it dissociates. This results in the presence of a continuous adsorption spectrum beyond a certain value $\mathrm{v}_{\mathrm{lim}}$ of v, so that the difference $\Delta_{\mathrm{v}}^{\mathrm{v+1}}E=0$, which gives us the value:

$$\mathrm{v}_{\mathrm{lim}} = \frac{1}{2x}-1 = \frac{2E_{\infty}}{\mathrm{h}\nu_{\mathrm{vibr}}}-1 \qquad [4.30]$$

The energy of dissociation is therefore equal to the difference between the minimal energy (obtained for v = 0) and this limit energy denoted by $E_{\infty}$. Thus, we write:

$$D = E_{\infty} - E_{\mathrm{v=0}} \qquad [4.31]$$

The series of the different vibratory states, their limit and the dissociation of the molecule result, on the spectrum, in increasingly tightly grouped bands (Figure 4.2), up to the limit where we see a continuous spectrum which extends to very short wavelengths.

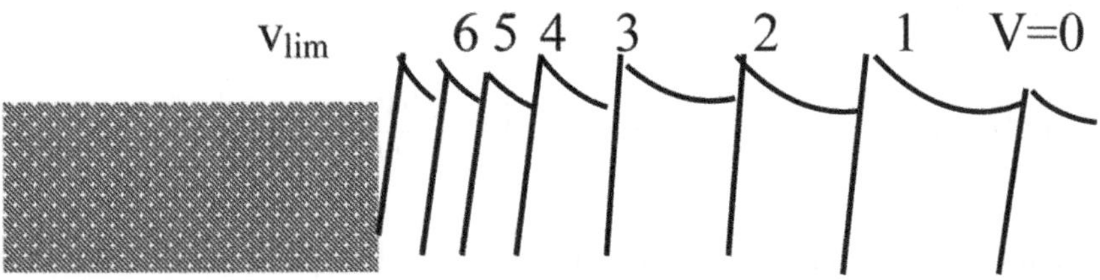


**Figure 4.2.** *Limit of a spectral series*

When we deduce the energies of dissociation using the spectral method, we are able to take account of the fact that the atoms may, following that dissociation, be in activated states, in which case it is useful to subtract the activation energy from the value given by the spectrum.

| Halogen | Energy of dissociation given by the spectrum | Activation energy | Spectroscopic dissociation energy | Thermal dissociation energy |
|---|---|---|---|---|
| Chlorine | 249 | 8 | 241 | 238.3 |
| Bromine | 232 | 43 | 189 | 192.3 |
| Iodine | 238 | 91 | 147 | 144.2 |

**Table 4.2.** *Comparison of the energies of dissociation of thermal origin and spectral origin for halogens (in kJ/mol) [EMS 51]*

For example, in the visible part of the spectrum, we find limits which correspond to 47.85, 51.07 and 49.95 nm for chlorine, bromine iodine respectively. From this, we deduce the values of ionization energy given by the first column in Table 4.2; however, this photochemical dissociation gives us one normal halogen atom and one activated atom, so we need to take off the energy of activation of that atom.

The second column of Table 4.2 gives the values of the energies of dissociation found by looking at the spectra for the three halogens. The third column in the table gives the energies of activation, which needs to be deducted for each of the three halogens.

We can then compare the values of the energy of dissociation thus obtained to the values found by thermal measurements. The comparison of the values shown in columns 4 and 5 of Table 4.2, in the case of halogens, exhibits a very close correspondence between the values obtained by the two methods.

#### 4.2.7.5. *Bond energies*

The concept of bond energy was introduced in 1920 by Fajans. It is at the heart of one of the earliest methods for evaluating unknown enthalpies of formation.

We saw earlier that the two O-H bonds in water each had its own energy. The energy contained in the O-H bond is defined as the mean of those two values, and more generally the mean of the enthalpies accompanying the reactions to break such an O-H bond in a large number of compounds. On the basis of the values we saw in the previous section, this bond energy would be $\varepsilon_{\text{O-H}} = 462.3 \text{ kJ.mole}^{-1}$. This value is, in fact, half of the standard enthalpy of synthesis of water, which is why knowledge of these enthalpies of synthesis is valuable.

Thus, it has been possible to attribute a bond energy to a large number of bonds – particularly in the domain of organic chemistry. Table 4.3 shows a number of such values, expressed in kJ/mole.

| **Bond** | **Energy** | **Bond** | **Energy** | **Bond** | **Energy** |
|---|---|---|---|---|---|
| H-H | 435.89 | C-C | 345.6 | C-O | 357.7 |
| C-H | 413.0 | C=C | 610.0 | C=O | 748 |
| O-H | 462.8 | $C \equiv C$ | 815.1 | C-N | 304.6 |
| Cl-H | 431.4 | C-Cl | 339 | $C \equiv N$ | 889.5 |

**Table 4.3.** *A few values of bond energies (kJ.mol$^{-1}$)*

These bond energies can be obtained on the basis of means of the energies of dissociation of the bonds. Another method of determining these bond energies is to use the energies of combustion. For instance, let us look at the C-H bond; we start with the reaction of combustion of methane by the reaction:

$$CH_4(\text{gas}) + O_2(\text{gas}) = CO_2(\text{gas}) + H_2O(\text{gas}) \qquad [4R.4]$$

$$(\Delta_{R4} h^0_{298} = -890.36 \text{ kJ.mol}^{-1})$$

Then, the inverse reactions of the reactions of formation of carbon dioxide and water, which are:

$$CO_2(\text{gas}) = C(\text{graphite}) + O_2(\text{gas}) \qquad [4R.5]$$

$$\Delta_{R5} h^0_{298} = 393.51 \text{ kJ.mol}^{-1}$$

$$2H_2O(\text{gas}) = 2H_2(\text{gas}) + O_2(\text{gas}) \qquad [4R.6]$$

$$\Delta_{R6}h^0_{298} = 571{,}70 \text{ kJ.mol}^{-1}$$ [2]

The inverse reaction of the reaction of dissociation of hydrogen is:

$$H_2(\text{gas}) = 2H(\text{gas}) \quad [4R.7]$$

$$\Delta_{R7}h^0_{298} = 871.86 \text{ kJ.mol}^{-1}$$

Finally, we have the reaction of sublimation of graphite:

$$C(\text{graphite}) = C(\text{gas}) \quad [4R.8]$$

$$\Delta_{R8}h^0_{298} = 716.682 \text{ kJ.mol}^{-1}$$

The sum of reactions [4R.4] to [4R.8] gives us the following for the dissociation of methane:

$$CH_4(\text{gas}) = C(\text{gas}) + 4\ H(\text{gas}) \quad [4R.9]$$

This sum gives us:

$$\Delta_{R9}h^0_{298} = 1663.39 \text{kJ.mol}^{-1}$$

This sum can be considered as being four times the energy of the C-H bond, which therefore is:

$$\varepsilon_{\text{C-H}} = 415.85 \text{ kJ.mol}^{-1}$$

Note that many bond energies use the sublimation of graphite, for which, as pointed out above, it is difficult to obtain a value. This accounts for the sometimes-divergent values noted between different sources of values of those bond energies – the divergence stems from a difference in the choice of the sublimation energy of graphite. Thus, the value of the C-H bond is given as 388.74 kJ/mol by the *Handbook der Physik*, based on a value of 627 kJ/mol for the sublimation of graphite, whereas Pauling, in 1945, gave the value 364.91kJ/mol, based on graphite sublimation at 578.74 kJ/mol, and in 1948, Gerö and Valatin gave the value of 405.46 kJ/mol, based on 709.35 kJ/mol for graphite sublimation. The value calculated above is 716.68 kJ/mol, and it is this reading which we use as the basis for the values reported in Table 4.4.

NOTE 4.3.– it is important not to confuse the bond energy, which is an average value, with the energy of dissociation of that bond, which is the energy needed to break the bond in a given compound. For example, the energy of dissociation of the C-H bond in methane is different depending on whether we are talking about breaking the first C-H bond or breaking the second, with the first already having been broken.

#### 4.2.7.6. *Estimation of the enthalpies of formation on the basis of the bond energies*

The enthalpy of formation of a compound can quickly be evaluated on the basis of the bond energies; we merely need to add together all of the bond energies in the compound, which gives us the opposite of the enthalpy of the reaction by which the compound is synthesized. Based on that enthalpy of synthesis, we are able to work out the standard energy of formation (see section 4.2.7.1).

## 4.3. Reaction entropies

For the enthalpies, we have also managed to assign to each compound a standard enthalpy by the standard enthalpies of formation, by arbitrarily giving the value of zero to the enthalpy of formation of the simple substances at any temperature. In the case of entropies, we shall construct a scale by deciding that for all solid substances, the entropy has the same value at a certain temperature. Thus, we shall define an absolute scale of the entropies.

### 4.3.1. *Planck's hypothesis – calculation of the calorimetric entropies*

In 1911, Planck hypothesized that the entropies of all crystalline substances tend toward zero as their temperature tends toward zero. Thus, we write:

$$s_0^T \rightarrow 0 \text{ as } T \rightarrow 0 \qquad [4.32]$$

We deduce the same property for the standard entropy associated with the reaction:

$$\Delta_r s_0^0 = \sum \nu_k s_{0(k)}^0 = 0 \qquad [4.33]$$

This property has been dubbed the third law of thermodynamics.

We can then use relation [4.8] to calculate the standard entropy at any temperature, by making $\Delta_r s_0^0 = 0$ and $T_0 = 0$ K, which gives us:

$$\Delta_r S_T = \sum_{\varphi} \int_{T^{(\Delta\varphi)}}^{T} \frac{\Delta_r C_P^{(\varphi)}}{T} \mathrm{d}T + \sum_{\Delta\varphi} \nu_k \frac{\Delta_{\Delta\varphi} H_k}{T^{(\Delta\varphi)}} \qquad [4.34]$$

As the specific heat capacities and the latent heats of state change are measured by calorimetry, we use the term *calorimetric standard entropy* to speak of the value calculated by relation [4.34].

In order to verify the values obtained, we can also determine the values of the standard entropy when we know the equilibrium constant at the desired temperature and the standard enthalpy of reaction at the same temperature. Then by combining expressions [4.1] and [4.2], we obtain:

$$\Delta_r s_T^0 = -\mathrm{R} \ln K_P + \frac{\Delta_r h_T^0}{T} \qquad [4.35]$$

Comparisons have been carried out between standards calorimetric values and values measured on the basis of relation [4.35]. Table 4.4 shows a few results. We can say that, in spite of the major difficulty in measuring equilibrium constants, the comparison is fairly satisfactory, which gives a good verification of Planck's hypothesis.

| Reaction | Based on the equilibria | Calorimetric |
|---|---|---|
| C(graphite) + ½ $O_2$ = $CO_2$ | 89.3 | 84.02 |
| Hg + ½ $O_2$ = HgO | -109.10 | -108.26 |
| CO + $H_2O$ (liq) = $CO_2$ | -126.55±6.3 | -125.82±8.4 |
| C(graphite) + ½ $O_2$ = $HCO_2H$ | -206.07±12.5 | -202.73±6.3 |
| 1/2 Pb + AgI = 1/2 $PbI_2$ + Ag | -16.72 | -14.00 |

**Table 4.4.** *Comparisons of the standard entropies of reactions at 25°C, $\Delta_r S_{298}^0$ (in Jmol$^{-1}$.deg$^{-1}$), measured on the basis of the equilibrium constants, by calorimetry [EMS 51]*

### 4.3.2. *Spectroscopic determination of entropies – absolute entropies*

Statistical thermodynamics enables us to express the entropy as a function of the canonical partition function $Z_C$ (relation [A2.39], see Appendix 2). This partition function is expressed by relation [A2.36], on the basis of the molecular partition functions. These molecular partition functions are expressed, in relation [A2.21], by the partition functions of translation, vibration and rotation. These are calculated on the basis of the molecule mass and relation [A2.26] for a perfect gas, the vibration frequencies (relation [A2.30]) of its bonds and of its moments of inertia (expression [A2.29]). These data are determined by studying the spectra of the molecules – particularly the absorption spectra in the infrared. Hence, at least for simple molecules, we are able to calculate an *absolute* value for the entropy – i.e. with no frame of reference, and in particular without the aid of Planck's hypothesis.

### 4.3.3. *The third law*

We have given an initial approximation of the third law on the basis of Planck's hypothesis. We shall come back to this formula after comparing the values of the calorimetric standard entropies (extracted from relation [4.34]) with the absolute entropies measured with spectroscopic data.

#### 4.3.3.1. *Comparison between calorimetric and spectroscopic sources – residual entropy*

Table 4.5, for a number of substances, gives the values found by calorimetry, calculated using Planck's hypothesis, and the absolute values found by spectroscopic measurements, independent of Planck's hypothesis. We can see two distinct categories of molecules:

– firstly, we have those, such as nitrogen and oxygen, with which there is a satisfactory correspondence between the values found by the two techniques; and

– secondly, there are those, such as the other molecules shown in Table 4.5, where the spectroscopic value is consistently higher than the calorimetric value. The fourth column in Table 4.5 expresses these differences.

| Substance | Calorimetry | Spectroscopy | Experimental difference | Theoretical difference |
|---|---|---|---|---|
| $N_2$ | 191.86 | 191.40 | -0.46 | 0 |
| $O_2$ | 205.24 | 204.95 | -0.29 | 0 |
| $H_2$ | 124.15 | 130.54 | 6.40 | 0.75Rln3 = 6.86 |
| $D_2$ | 141.70 | 144.71 | 3.01 | 0.5Rln3 = 2.88 |
| CO | 193.12 | 197.80 | 4.68 | Rln2 = 5.73 |
| NO | 215.02 | 219.78 | 4.77 | Rln2 = 5.73 |
| $H_2O$ | 185.26 | 188.69 | 3 .43 | Rln(3/2) = 3.43 |

**Table 4.5.** *Calorimetric and spectral values of the entropies at 298K for certain substances (expressed in J.mol$^{-1}$.deg$^{-1}$), and zero entropies, found experimentally and calculated [EMS 51]*

Observers have deduced from these data that, with certain molecules, there is a difference between the two series, which in fact expresses a non-null, positive value of the entropy in these solid substances at the temperature of absolute zero. This is known as the *residual entropy* or *zero entropy*.

Evidently, such a result means that we need to review Planck's hypothesis, and therefore the way in which the third law of thermodynamics is stated.

#### 4.3.3.2. *Statement of the third law of thermodynamics*

The currently-accepted statement of the third law of thermodynamics is as follows: the entropy of any substance tends toward a finite value at the temperature of absolute zero, and that value may be equal to zero in a number of cases.

Some authors specify that substances at absolute zero must be crystallized. However, it seems that, in the case of helium, we can never achieve the crystallized state; instead the state at absolute zero is a liquid form, but it has been demonstrated that its entropy tends toward zero.

The third law means that calorimetric measurements of entropy are not reliable, because we do not know the residual value at 0 K. Unfortunately, the spectral method is only applicable for molecules whose spectra are sufficiently well interpreted – i.e. are not overly complex. Hence, we only have an accurate measurement of the absolute entropy for a limited number of substances.

NOTE 4.4.– as noted in section 4.2.5.3, we can treat the enthalpy of formation and the pure-substance enthalpy as one and the same thing, which is expressed by the following double relation, for a given substance:

$$\Delta_f h_T^0 = H_T^0 = h_T^0 \tag{4.36}$$

The same is no longer true for the absolute entropies, which are not entropies of formation. Thus, we have:

$$\Delta_f s_T^0 = S_T^0 \neq s_T^0 \tag{4.37}$$

The entropy of formation is calculated on the basis of the absolute entropies of the simple substances and the stoichiometric coefficients of the formation reaction.

Hence, the same is true for the Gibbs energies:

$$\Delta_f g_T^0 = G_T^0 \neq g_T^0 \tag{4.38}$$

#### 4.3.3.3. *Residual entropy of crystalline substances*

The existence of a residual entropy at the temperature of 0 K means, by application of Boltzmann's equation, that the number of complexions $\Omega$ is not necessarily equal to 1 at that temperature. Thus, absolute order is not always reached, and the value of the number of complexions must be able to be greater than 1 (never less than 1, of course, which would be absurd. This explains why the residual entropy value is always positive). By quantitative study, we have been able to quantify the initial state of disorder and evaluate the residual entropy. To exhibit the method, we shall consider a crystal of carbon monoxide, in which the two oxygen and carbon atoms are differentiated, but are nonetheless very similar. In particular, they have the same weak electric dipole moment.

If we examine a crystallographic assay of the lattice of crystallized CO, the molecules may have either the configuration "CO" or the configuration

"OC". If these two arrangements are practically equivalent in terms of energy, this means that the number of complexions at absolute zero for a set of N molecules (n moles) will be $\Omega = 2^N$, and thus the residual entropy of CO will be:

$$s_0^0 = k_B \ln 2^N = N k_B \ln 2 = n R \ln 2 \qquad [4.39]$$

Generally speaking, if $s$ represents the number of equivalent configurations at absolute zero, the molar residual entropy will be:

$$s_0^0 = R \ln s \qquad [4.40]$$

The last column in Table 4.5 shows the values calculated for a number of molecules, and we see a very close correspondence with the observed values, shown in the fourth column.

NOTE 4.5.– Up until now, we have defined an entropy known as the absolute entropy, because it is characterized by a number of complexions equal to 1. In fact, in that number, we have only taken account of the states of the nucleus, the electrons and the atoms. There is nothing to suggest that, were we to take account of the states of the particles internal to the nucleus such as protons, neutrons or other nuclear subatomic particles, we would actually obtain the same "absolute" entropy. In fact, our absolute entropy can be qualified as *chemical*: strictly speaking, it remains a relative entropy value: that calculated by taking the number of complexions at the scale of the atom as equal to 1.

## 4.4. Specific heat capacities

The values of the specific heat capacities are found by two different kinds of experimental methods: calorimetric methods and spectroscopic techniques.

### 4.4.1. *Calorimetric measurements of the specific heat capacities*

Specific heat capacities found by calorimetry are usually represented as a function of the temperature, by functions in the form:

$$C_P = a + bT + cT^2 + dT^{-2} \qquad [4.41]$$

Certain coefficients may be equal to zero.

In order to establish such a function, we can either take measurements at different temperatures, each being taken at constant temperature, or take measurements at a temperature which varies in a linear fashion over time.

#### 4.4.1.1. *Isothermal measurements with a drop calorimeter*

In the isothermal method of determining the specific heat capacities, we use a drop calorimeter.

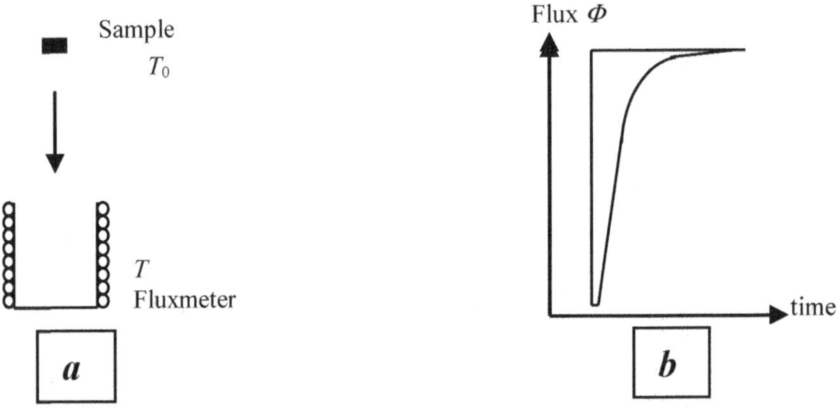


**Figure 4.3.** *Drop calorimeter*

We drop the sample (Figure 4.3(a)), taken to temperature $T_0$ (usually 25°C), into a thermal fluxmeter previously stabilized at the measuring temperature $T$, which is usually higher than $T_0$. The heat flux is recorded (Figure 4.3(b)) as a function of time, until the signal finally returns to its initial value. The area beneath the curve gives the energy which was needed to raise the temperature of the sample from temperature $T_0$ to the temperature $T$. We work from this value of the exchanged energy to find its value per mole by multiplying it by the ratio $m/M$ of the mass of the sample to the molar mass of the substance of which it is composed. The experiment is restarted at different temperatures $T$. From these measurements, we establish the coefficients of a law in the form:

$$H = aT + \frac{b}{2}T^2 + \frac{c}{3}T^3 - \frac{d}{T} \quad [4.42]$$

Using the following relation, which is accurate at constant pressure:

$$C_P = \frac{\mathrm{d}H}{\mathrm{d}T} \qquad [4.43]$$

we deduce the coefficients for relation [4.41].

#### 4.4.1.2. *Non-isothermal measurements by differential scanning calorimetry*

The sample is stabilized at a temperature $T_1$ in a low-inertia thermal fluxmeter (differential scanning calorimeter). Linear programming of the temperature with the constant slope $\mathrm{d}T/\mathrm{d}t = \beta$ is then imposed in the kiln of the instrument, up to the temperature of $T_2$. The thermal flux signal is then recorded between times $t_1$ and $t_2$ (Figure 4.4), which are sufficiently far apart for the flux signal to be stable. The experimental temperature $T$ is taken as the mean of the two temperatures $T_1$ and $T_2$.

$c_P$ and $C_P$ respectively denote the specific mass heat of the sample and its molar specific heat capacity at constant pressure; $m$ is its mass and $M$ the molar mass of the substance in question. We can write the following for the flux:

$$\Phi = \frac{dH}{dT} = mc_P \frac{dT}{dt} = \frac{m}{M} C_P \beta \qquad [4.44]$$

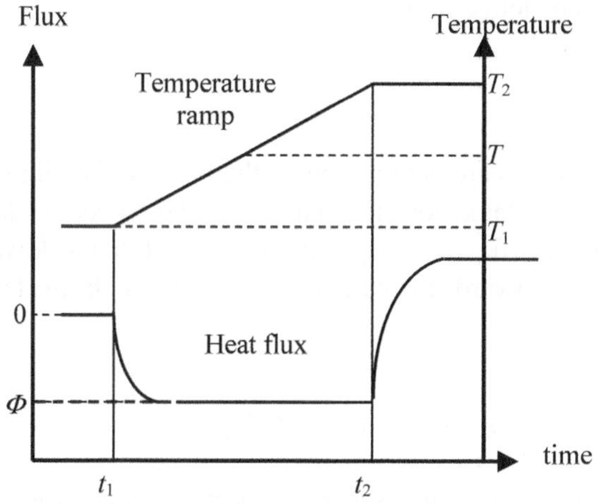


**Figure 4.4.** *Differential scanning calorimetry*

This gives us the value of $C_P$ deduced from the measured heat flux.

### 4.4.2. *Spectral measurements of the specific heat capacities*

The specific heat capacities are represented by functions derived from statistical models – e.g. for solids [SOU 15a], the Einstein function:

$$C_{V(v)} = 3\mathrm{R}\left(\frac{\Theta_E}{T}\right)^2 \frac{\exp\left(\frac{\Theta_E}{T}\right)}{\left[\exp\left(\frac{\Theta_E}{T}\right)-1\right]^2} \qquad [4.45]$$

or Debye function:

$$C_{V(v)} = 12\mathrm{R}\left(\frac{T}{\Theta_D}\right)^3 \int_0^{\Theta_D/T} \frac{x^3}{\exp(x)-1}\mathrm{d}x \qquad [4.46]$$

These relations respectively involve either the Einstein temperature:

$$\Theta_\mathrm{E} = \frac{\mathrm{h}\nu}{\mathrm{k_B}} \qquad [4.47]$$

or the Debye temperature $\Theta_D$, in which $\nu_D$ is the limit frequency of the distribution of the Debye frequencies:

$$\Theta_D = \frac{\mathrm{h}\nu_D}{\mathrm{k_B}} \qquad [4.48]$$

These temperatures are determined on the basis of the vibration frequencies of the bonds calculated on the basis of the spectral data. Remember that the molar specific heat capacity of a solid tends toward zero as its temperature tends toward absolute zero (with the entropy tending toward a constant).

## 4.5. Experimental determination of the equilibrium constants

There are a number of ways to experimentally determine the value of an equilibrium constant. In fact, we need to be able to determine the fractional extent of the reaction at equilibrium, from which, when we know the initial state, we deduce the composition of the medium. If we know that composition, in the case of solutions, we deduce the activities of the substances, knowing the relation between the activity and the concentration of the solutions used.

There are numerous possible measuring methods, and ultimately they are the same as those used to track the kinetics of a reaction [SOU 11]. Traditionally, we distinguish between:

– chemical methods, in which we dose a sufficient number of components at the end of the reaction, but preventing the addition of a dosing reagent from disturbing the equilibrium by displacing it by capturing a component;

– physical methods, in which a physical measurement is characteristic of the extent of the reaction. These methods are easier to implement, but can only be used in relatively-simple cases.

In fact, the experimental difficulty is two-fold:

– we need to be certain that the *steady* state reached is indeed the equilibrium state, meaning that we achieve the same composition of the medium regardless of which side of the reaction we begin on;

– the slowness or the rapidity of the reactions often prevent us from covering a sufficiently broad temperature range. The use of a catalyst enables us to use broader temperature domains without altering the equilibrium properties.

A particularly elegant method to determine the equilibrium constants is, when possible, to use the measurement of electromotive forces (emf) of galvanic elements or electrochemical cells.

We know that an electrochemical cell works by way of a chemical reaction, transforming chemical energy into electrical energy, and that the standard emf $E^0$ of the cell or standard potential (i.e. the emf measured at standard pressure with zero current – in an open circuit) at a given temperature is linked to the standard Gibbs energy associated with the reaction at the same temperature by the expression:

$$\Delta_r g^0 = -E^0 z\mathfrak{F} \qquad [4.49]$$

In that expression, $\mathcal{F}$ is the faraday, which is 96,500 coulombs and $z$ is the number of electrons exchanged in the stack reaction, which is a redox reaction. Thus, by measuring the standard potential, we are able to find the standard Gibbs energy.

The system is given reversibility by counterbalancing the measuring cell with an electromotive force set up in opposition, in such a way that the current is minimal. A slight increase or decrease in the strength of the opposing emf changes the direction in which the reaction takes place.

The main precaution is to ensure that the stack reaction is really that which has been studied, and that it is not disturbed by secondary reactions.

## 4.6. Calculation of the equilibrium constants on the basis of other thermodynamic data

It is tricky to experimentally determine the equilibrium conditions, so as much as possible, we use the calculation method to determine the values of the equilibrium constants (or standard Gibbs energies associated with the reactions). Those calculations must restrict the number of experiments as far as possible, or eliminate them altogether.

There are five calculation methods which we can envisage, depending on the data available to us:

– method 1: we know the enthalpy and the standard entropy of reaction at the desired temperature;

– method 2: we know the standard enthalpy and the standard entropy at a specific temperature, and the molar specific heat capacities of components at constant pressure as a function of the temperature;

– method 3: we know a value of the equilibrium constant at a given temperature, the variations of the specific heat capacities at constant pressure of the components with varying temperature, and a value of the standard enthalpy associated with the reaction at a given temperature;

– method 4: we know two values, at two temperatures, of the equilibrium constant and the molar specific heat capacities at constant pressure of the components as a function of the temperature:

– method 5: we know other, judiciously-chosen equilibrium constants.

We shall examine each of these five methods in turn.

### 4.6.1. *Calculation of the equilibrium constants – method 1*

We suppose the following to be known:

– the standard enthalpy of the reaction at the desired temperature;

– the standard entropy of the reaction at the desired temperature.

This is the simplest case. Indeed, we can immediately calculate the equilibrium constant at the same temperature, by applying relations [4.1] and [4.2].

### 4.6.2. *Calculation of the equilibrium constants – method 2*

We know:

– the standard enthalpy at a temperature $T_0$;

– the standard entropy at a temperature $T_0$;

– the variations of the specific heat capacities at constant pressure of the components as a function of the temperature and the enthalpies of the state changes.

Using a relation such as [4.9], we find the value of the equilibrium constant at the chosen temperature.

This method enables us to calculate an *a priori* equilibrium constant without ever needing to take any measurements in equilibrium conditions. Thus, it will also enable us to find the constant for an equilibrium which is not practically realizable.

### 4.6.3. *Calculation of the equilibrium constants – method 3*

We know:

– a value of the equilibrium constant at a temperature $T_0$;

– the variations of the specific heat capacities at constant pressure of the components with the temperature and the enthalpies of the state changes; and

– a value of the standard entropy at the temperature $T_0$.

By applying relation [4.10], we obtain the equilibrium constant sought at the chosen temperature $T$.

### 4.6.4. *Calculation of the equilibrium constants – method 4*

We know:

– two values of the equilibrium constant of a reaction at two temperatures $T_1$ and $T_2$;

– the variations of the molar specific heat capacities of the components at constant pressure as a function of la temperature and the state-change enthalpies.

We can then calculate the equilibrium constant at any other temperature $T_3$.

By combining the equations obtained using equation [4.9] for each of the temperatures $T_1$ and $T_2$, we can extrapolate the value of the standard enthalpy at temperature $T_0$, $\Delta_r h_0^0$ and that of the standard entropy at the same temperature $\Delta_r s_0^0$. Then, we need only use expression [4.9] for the chosen temperature $T_3$.

### 4.6.5. *Calculation of the equilibrium constants – method 5*

We can calculate the equilibrium constant on the basis of other known equilibrium constants, which are appropriate at the chosen temperature. We use our knowledge of the equilibrium constants of other reactions, chosen in such a way that a linear combination of their balance equations gives us the balance equation of the reaction at hand. The unknown standard Gibbs energy is deduced from the known standard Gibbs energies by the same linear combination.

Let us look at the example of the reaction of synthesis of water vapor. The balance equation is written as:

$$CO_2 + H_2 = CO + H_2O \qquad [4R.10]$$

We can see that if we consider the two reactions:

– synthesis of water, given by:

$$H_2 + ½\,O_2 = H_2O \qquad [4R.11]$$

– and combustion of carbon monoxide, by:

$$CO + ½\,O_2 = CO_2 \qquad [4R.12]$$

the balance equation of the reaction in question is the difference between the balance equations for reactions [4R.11] and [4R.12]. Thus, we have:

[4R.10]= [4R.11]– [4R.12]

Let us apply the same linear combination to the standard Gibbs energies, and thus:

$$\Delta_{10}g^0 = \Delta_{11}g^0 - \Delta_{12}g^0$$

The application of relation [4.2] immediately gives us the relation between the sought equilibrium constant and the known equilibrium constants:

$$K_{10}^{(1)} = K_{11}^{(1)} / K_{12}^{(1)}$$

Hence, if $K_{11}^{(1)}$ and $K_{12}^{(1)}$ are known at the desired temperature, we can easily calculate $K_{10}^{(1)}$.

Ultimately, the solving of all problems of determination of the equilibrium constants (or the standard Gibbs energies) relies on our knowing:

– the standard enthalpy at a chosen temperature;

– the standard entropy at the same temperature;

– the variations of the molar specific heat capacities as a function of the temperature.

We have seen the advantage of going into greater detail about the properties and methods for measuring or calculating those values in sections 4.2, 4.3 and 4.4.

## 4.7. Determination of the equilibrium constants on the basis of spectral data and statistical thermodynamics

In Appendix 2 (section A2.5), we recapped the relations that exist between the equilibrium constant for different types of reaction, the molecular partition functions and the difference between the fundamental energies at the temperature of 0 K of the substances involved in the reaction. These partition functions and this energy difference can be calculated on the basis of the vibration frequencies of the molecules, their moments of inertia and their masses. Those data can be accessed on the basis of the absorption spectra, essentially in the infrared. Thus, it is possible to calculate the equilibrium constants *a priori*.

We can show that the relations obtained are valid if the polycomponent phases exhibit perfect behavior in the sense of the solutions. In the case of gases, those relations are applicable to real gases, provided the mixture of gases is perfect in the sense of the solution. Thus, the relations are applicable when the Lewis approximation is satisfactory (see relation [3.5]). In the case of solutions with non-perfect behavior, to the difference between the fundamental energies we must add the energy (or enthalpy) of mixing, which can be obtained either on the basis of the measurements or by modeling the interactions between the molecules in the solution.

## 4.8. Thermodynamic tables and databanks

There are a wide variety of thermodynamic tables. Some are very broad, and others cover a particular family of products. These tables can be classified into:

– two families relative to the date when they were established;

– two families relative to their degree of accuracy and finesse.

Depending on the dates at which they were established, recent tables are set apart from older ones by the units employed and the choice of the pressure in the standard state.

The recent tables are established in the international system (SI), and choose the Joule as the unit of energy and 1 bar as the pressure of the standard state.

The old tables usually use the kcal/mol for enthalpies and the $cal.mol^{-1}K^{-1}$ (or clausius, which is identical) for entropies. The standard state is chosen for the pressure of 1 atmosphere. Thus, we have the following conversions to switch to the international system:

– for the enthalpies, multiply by $4.184\times10^3$ to switch from kilocalories per mole to joules per mole. The effect of pressure is negligible;

– for the entropies, it is helpful to multiply the values by 4.184 to switch from clausius or $cal.mol^{-1}K^{-1}$ to $J.mol^{-1}K^{-1}$, and add the term $R.\ln 1.01325 = 0.1094$, to take account of the change of pressure for the standard state;

– for the Gibbs energies, we multiply the ancient value by 4.184 to switch from calories to joules, and we add the term $RT.\ln 1.01325$ to take account of the pressure change for the standard state.

In terms of the detailed data, we distinguish the basic tables and the detailed tables.

The basic tables, for each substance in a given physical state (crystalline solid in a given variety, liquid or gas), give us:

– the standard enthalpy of formation at the temperature of 298.15 K (25°C);

– the standard entropy at the temperature of 298.15 K (25°C);

– an equation showing the variation of the molar specific heat capacities at constant pressure, at standard pressure, with temperature and the temperature range of its validity; and

– the temperatures of state change and the enthalpies associated with those transformations (sometimes called the latent heat of state change) at that temperature.

Certain tables are supplemented by the standard Gibbs energy of formation at 298 K and the logarithm of the equilibrium constant of the reaction of formation of the compound at the temperature of 25°C.

The detailed tables, for certain temperature values, generally at constant intervals, and a specific physical state of the substance, give us the following data:

– the value of the specific heat capacity standard at constant pressure;

– the standard entropy;

– the value $-\left(g_T^0 - h_{298}^0\right)/T$ which is, in fact, the difference between the standard entropy at temperature $T$ and the standard entropy at the reference temperature;

– the difference $h_T^0 - h_{298}^0$ between the standard enthalpy of formation at temperature $T$ and that at the reference temperature;

– the standard enthalpy of formation at temperature $T$: $h_T^0$;

– the standard Gibbs energy of formation at temperature $T$: $g_T^0$;

– the logarithm of the equilibrium constant for the formation at temperature $T$: $\ln K_f^{(1)}$.

In fact, it is imperative, before using a thermodynamic table or a databank, to carefully read the accompanying notice, which usually specifies the definitions of the values given and the units used. Indeed, the symbols of the different values are not always the same from one table to another, and it is possible for confusion to arise. We also need to beware of confusing the decimal logarithm and the Napierian logarithm. The values given may be molar, gravimetric or volumetric.

When the data are provided for specific values of the temperatures, linear interpolations between two values or parabolic interpolations with three values can be used to obtain non-tabulated data.

## 4.9. Estimation of thermodynamic data

In practice, we can see that we actually have very scant data on the standard enthalpies and standard entropies of compounds, found by reliable measurements, and that those data are not known for a large number of compounds. Hence, users have sought to estimate, with the greatest possible degree of accuracy, values for those data when they have not had values gleaned from calorimetric experimentation or the examination of the adsorption spectrum.

### 4.9.1. *Method of evaluating the energies of dissociation by spectroscopy*

We have seen (section 4.2.7.4) the method for determining the bond energies on the basis of the limit of the continuous absorption spectrum. This limit, though, is often fairly difficult to observe directly. We shall see, however, that we can estimate the energy of dissociation by calculation, by determining the law which gives the position of the successive vibration bands.

Let us look again at the approximate expression [4.25] for the absorption energy corresponding to a spectral line for the value v of the level parameter. We shall apply that relation for the first two lines in the spectral series characterized by the transitions $v = 1 \rightarrow v = 2$ and $v = 2 \rightarrow v = 3$. We obtain:

$$\Delta_1^2 E = h\nu_{vibr}\left[1 - \frac{h\nu_{vibr}}{E_\infty}\right] \qquad [4.50]$$

and:

$$\Delta_2^3 E = h\nu_{vibr}\left[1 - 6\frac{h\nu_{vibr}}{E_\infty}\right] \qquad [4.51]$$

By solving the system composed of these two equations [4.50] and [4.51], we obtain an estimated value of the energy $E_\infty$ and of the vibration frequency $\nu_{vibr}$. The energy of dissociation is then calculated by the relation:

$$D = E_\infty - \frac{h\nu_{vibr}}{2} \qquad [4.52]$$

In the evaluation of the enthalpy of formation, we can then treat the energy of dissociation and the bond energy as identical.

### 4.9.2. *Group contribution methods*

We have shown how the concept of bond energy could facilitate an estimation of the standard enthalpies of formation on the basis of the molecular structure. In practice, we are evaluating not the enthalpy but the standard Gibbs energy at a given temperature.

Different evaluation methods exist, and are generally applied to a family of products. Two important categories are frequently used: that applied to metals and liquid metal alloys and that used for organic products.

For this latter family, most methods for evaluating the thermodynamic data are inspired by what we have seen for bond energies, based on the structure of the molecule. The most significant methods are based on what we call the functional-group substitution method. They are founded on a certain number of observations. For example:

– if a methyl group ($CH_3$) is substituted for a hydrogen atom attached to a carbon atom, we note an increase in the standard Gibbs energy of formation at 298 K by around 7948 J/mol;

– if a hydroxyl group (OH) is substituted for a hydrogen atom attached to a carbon atom, we see a decrease of the standard Gibbs energy of formation at 298 K, by around 142 120 J/mol.

Thus, when we know that the standard Gibbs energy of formation of ethane at 298 K is -32,604J/mol. For propane, which results from the substitution of a hydrogen atom by a methyl group on a carbon atom from ethane, we then calculate a standard Gibbs energy of formation at 298 K of:

$$-32{,}604 + 7945 = -34{,}602 \text{ J/mol}$$

knowing that the experimental value is –23,408 J/mol.

Similarly, for ethyl alcohol, we can easily calculate a standard Gibbs energy of formation at 298 K:

$$-32{,}604 - 142{,}120 = -174{,}724 \text{ J/mol}$$

for a real value of –168,038 J/mol.

Expanding on these observations, different tables have been drawn up, giving the contributions of various substitutions on the thermodynamic values. A series of particularly complete tables are given by Andersen,

Berger and Watson. These tables, for a large number of cases, give the result of a substitution on:

– the standard enthalpy of formation at 298 K;

– the standard entropy at 298 K;

– the coefficients *a*, *b* and *c* of the first three terms in the limited expansion of the specific heat capacity with temperature.

Table 4.6 gives the example of such a table for the substitution of a methyl group for a hydrogen atom on methane and cyclopentane. The substance is supposed to be in the state of a perfect gas. Note that for the second substitution on the latter hydrocarbon, a distinction is made as to the degree of substitution of the carrier carbon atom and the one most closely adjacent to it. Each of those carbons is characterized by a number giving its degree of substitution, as shown by Table 4.7.

| **Base group** | **Substitutions** | $\Delta\left(H^0_{298}\right)$ | $\Delta\left(S^0_{298}\right)$ | Δ *(a)* | Δ *(b)* | Δ *(c)* |
|---|---|---|---|---|---|---|
| Methane | | -9830 | 13.30 | 9.92 | 103.81 | 43.51 |
| Cyclopentane | 1st substitution | -34,430 | 49.25 | 8.14 | 68.24 | 25.18 |
| | 2nd substitution | | | | | |
| | 1-1 | -26,610 | 17.15 | -6.02 | 116.36 | -55.56 |
| | 1-2 cis | -17,870 | 24.02 | -3.04 | 110.46 | -54.22 |
| | 1-1 trans | -25,020 | 24.69 | -2.47 | 107.57 | -52.13 |
| | 1-3 cis | -25,020 | 24.69 | -2.47 | 107.57 | -52.13 |
| | 1-3 trans | -25,020 | 24.69 | -2.47 | 107.57 | -52.13 |

**Table 4.6.** *Examples of contributions of substitutions of a methyl group with one hydrogen on the methane and cyclopentane bases*

| Number of types | Nature |
|---|---|
| 1 | $-CH_3$ |
| 2 | $>CH_2$ |
| 3 | $>CH-$ |
| 4 | $>C<$ |
| 9 | Carbon in an aromatic ring |

**Table 4.7.** *Characterized carbon atoms*

Besides the substitutions on the five base molecules in Table 4.8, these tables also show the effects of the introduction of multiple bonds.

| Base groups | $H^0_{298}$ | $S^0_{298}$ | *a* | *b* | *c* |
|---|---|---|---|---|---|
| Methane | -74,850 | 186.19 | 16.69 | 65.61 | -9.96 |
| Cyclopentane | -77,240 | 292.88 | -41.92 | 473.71 | -182.59 |
| Cyclohexane | -123,140 | 298.24 | -52.22 | 599.78 | -230.91 |
| Benzene | 82,930 | 269.20 | -22.51 | 402.54 | -171.42 |
| Naphthalene | 150,960 | 335.64 | -28.41 | 623.25 | -268.91 |

**Table 4.8.** *Thermodynamic data on the base groups in the group-substitution evaluation method*

We must not lose sight of the fact that, whichever method is used, these are evaluation methods which are incapable of delivering the same degree of precision as the values found by experimentation. In spite of their imperfections, numerous computer-based systems for calculating complex equilibria have databases which, for want of exact values, use such evaluation methods as these to find unknown thermodynamic values.

## 4.10. Thermodynamic calculations for complex systems

A complex system is a system subject simultaneously to several equilibria, which may be physical equilibria of state change and chemical equilibria between neutral molecules or ions. All these types of equilibrium may be encountered in the same system.

To calculate a system is to give, in specific conditions – i.e. of temperature, pressure, concentrations of certain species, and initial conditions – the nature and composition of the phases in equilibrium.

The calculation requires a clear definition of the system, and the result obtained is valid only in the strict context of that definition.

### 4.10.1. *Definition of the system*

The system is defined by the exhaustive list of the compounds and phases which the operator takes into consideration, and the choice of constraints s/he imposes, such as the temperature, the pressure or the domains of temperature, pressure, the concentrations or the initial quantities.

This definition is highly important, as no computation method can show the existence of a compound in a specific phase which the operator has not chosen in the definition list.

The conditions of evolution imposed on the system must be defined:

– isothermal or adiabatic;

– isobaric or isochoric (constant volume);

– closed or open system with a specific chemical potential (or partial pressure) imposed for a component.

The definition of the system also specifies a list of the independent reactions taken into account, including inter-phase reactions. It is not necessary for those reactions to actually take place, because we are dealing here with equilibrium calculations. Yet it must be borne in mind that such equilibria impose specific relations between the molar Gibbs energies of their components. Certain automated computation methods, which are specialized for certain types of systems, reveal only those equilibria.

### 4.10.2. *Output mode – graphical representation*

The results of the calculation can be presented in one of two forms:

– the descriptive form, if all the conditions were fixed, presents the values of the enthalpy and entropy of the system, its specific heat capacity at constant volume and pressure (the Gibbs energy of the system must be null, or in practice, extremely slight), the list of phases and their compositions, etc.; or

– a graphical output, if one of the constraints was only to remain within a specific range.

In fact, there are two main families of computation methods:

– the equilibrium-constant-based method;

– the method of minimization of the Gibbs energy, which is most commonly used for automated calculations.

### 4.10.3. *Calculation method based on the equilibrium constants*

By writing the law of mass action for each of the independent transformations and the conservation of species, we obtain analytical expressions (sometimes making certain approximations to simplify the calculations) for the molar fractions (or partial pressures) of each of the components as a function of the equilibrium variables, which may include molar fractions (or partial pressures of certain components) and equilibrium constants.

It must be acknowledged that the method quickly becomes highly tricky, and is rarely possible without a certain degree of approximation. Simple software tools are able to perform a quick calculation, although it may not be hugely accurate.

A variant of the method is particularly widely used for quickly performing the calculations for ionic solutions: the predominant reaction method, which is based on the concept of thermodynamically-equivalent systems. The method is usually described when studying these ionic reactions, for which it is very often employed.

Certain automated calculation software packages use the equilibrium-constants method.

### 4.10.4. *Method of minimization of the Gibbs energy function*

Modern numerical methods are built upon the fact that, at equilibrium, the Gibbs energy of the system should be zero. In fact, it is extremely difficult, *a priori*, to obtain a value of zero, so calculation software tends to use the minimization of the Gibbs energy function.

Consider a system containing $n$ chemical elements (in the sense of the periodic classification) divided between $N$ components. The Gibbs energy of the system will be given by the weighted sum of the molar fractions of the Gibbs energies of each of the $N$ components, as follows:

$$G = \sum_{j=1}^{N} n_j G_j \qquad [4.53]$$

The quantities $n_j$ are such that:

$$\sum_{j=1}^{N} n_j = n_t \qquad [4.54]$$

For each component, its molar Gibbs energy is such that:

– if it belongs to a gaseous phase, it is given by:

$$G_j = g_j^* + RT \ln \frac{f_j}{P^0} \qquad [4.55]$$

– if it belongs to a liquid or solid solution, we write:

$$G_j = g_j^0 + RT \ln \gamma_j x_j \qquad [4.56]$$

– if it is pure of phase, we have:

$$G_j = g_j^0 \qquad [4.57]$$

Note that, of the $N$ components of the system, the same chemical species present in several phases is counted as one component in each of those phases.

In the case of a very abundant solvent which does not participate as a component in an equilibrium, and therefore whose Gibbs energy is practically constant, it is not included in the list of components. The consequence of this is that the sum of the molar fractions in that solvent shall not be equal to 1, but instead equal to the sum of the molar fractions of the solutes. This method avoids simultaneously using very high values of molar fractions (very close to 1) and very low values (very close to 0).

Thus, the problem at hand is finally to find the series of values of variables $x_1, x_2, \ldots x_N$ which minimize the function $G(x_1, x_2, \ldots x_N)$ subject to certain constraints. This is a classic mathematical problem which can be solved by a variety of algorithms – such as the Lagrange multiplier method with several variables.

The software tools available on the market usually come with their own databanks, and hence are often fairly specialized in a particular domain, such as petrochemistry, interphase equilibria in materials, thermodynamics of plasma projection, etc. The use of such software has become extremely commonplace; developers have invested a huge amount of effort in simplifying and clarifying in the ergonomics of the man–machine interfaces, in particular.

# APPENDICES

# Appendix 1

# Recap on the Reference States of Solutions

In this appendix, we recap a few concepts pertaining to solutions, and to reference states used in the definition of the activities.

## A1.1. Concentration and molar fraction

Consider a solution. We label the values relative to the solvent with the subscript 0, and those relative to the solutes with the subscript $s$. By definition of the molar fractions, we can write the ratio:

$$\frac{x_s}{x_0} = \frac{n_s}{n_0} \tag{A1.1}$$

However, the volume of the solution, as a function of the partial molar volumes of the components, is given by:

$$V = n_0 \overline{V_0} + \sum_s n_s \overline{V_s} \tag{A1.2}$$

Therefore, by using the relation defining the concentration, we find:

$$C_s = \frac{n_s}{V} = \frac{n_s}{n_0 \overline{V_0} + \sum_s n_s \overline{V_s}} = \frac{x_s}{x_0 \overline{V_0} + \sum_s x_s \overline{V_s}} \tag{A1.3}$$

Yet if the solution is sufficiently dilute, and if $v_0^0$ denotes the molar volume of the pure solvent, we can write:

$$\sum_s x_s \overline{V_s} << x_0 \overline{V_0} \cong \overline{V_0} \cong v_0^0 \qquad [A1.4]$$

Thus, the molarity can be written approximately as:

$$C_s \cong \frac{x_s}{v_0^0} \qquad [A1.5]$$

The molar volume depends on the temperature and pressure for a gas. It depends practically only on the temperature for a liquid or a solid.

## A1.2. Chemical potentials and activity coefficients

We know that the chemical potential of a component in a perfect solution is written in the form:

$$\mu_i = \mu_i^0 + \mathrm{R}T \ln x_i \qquad [A1.6]$$

For a real solution, Lewis refers to that expression of the chemical potential, attempting to preserve its form. In order to do this, he introduced an activity coefficient $\gamma_i$ as a function of the temperature, pressure, etc. and of the solution's composition, writing the chemical potential of a component of the solution in the form:

$$\mu_i = \mu_i^0 + \mathrm{R}T \ln \gamma_i x_i \qquad [A1.7]$$

The product of the activity coefficient and the molar fraction is known as the activity of the component in the solution. The activity in a real solution plays the same thermodynamic role as does the molar fraction in a perfect solution:

$$a_i = \gamma_i x_i \qquad [A1.8]$$

### A1.2.1. *Activity coefficients of reference states*

Expression [A1.7], in fact, introduces two functions: $\mu_i^0$, known as the reference chemical potential, and the activity coefficient $\gamma_i$. Therefore, this introduction requires another given. We choose the definition of the reference state by way of a convention.

In practice, we have seen the introduction of several conventions; three are frequently used, and we refer to these as conventions (I), (II) and (III). Each of these have a corresponding reference chemical potential. We denote those respectively as $\mu_i^{0(\mathrm{I})}$, $\mu_i^{0(\mathrm{II})}$ and $\mu_i^{0(\mathrm{III})}$. Each convention also has a corresponding activity coefficient, written as $\gamma_i^{(\mathrm{I})}$, $\gamma_i^{(\mathrm{II})}$ and $\gamma_i^{(\mathrm{III})}$, and an activity, written as $a_i^{(\mathrm{I})}$, $a_i^{(\mathrm{II})}$ and $a_i^{(\mathrm{III})}$. Whenever an expression is valid irrespective of the reference used, we omit the distinguishing sign (I), (II) or (III).

It goes without saying that the value of the chemical potential of a component of the solution does not depend on the reference chosen, but instead, as the chemical potential is the partial molar Gibbs energy, the Gibbs energy of the solution would depend on that reference, which would be absurd.

### A1.2.2. *Convention (I)*

Convention (I) takes as a reference the components in the pure state in the same state of aggregation as the solution (this is known as the pure-substance reference). In these conditions, the chemical potential of the reference state is the molar Gibbs energy of the pure component and the chemical potential is then written as:

$$\mu_i = g_i^0 + RT \ln \gamma_i^{(\mathrm{I})} x_i \quad \text{[A1.9]}$$

This referenceis primarily used when all the components of a solution play the same role, and in particular have comparable molar fractions. For example, this convention is chosen when a given covers a broad spectrum of compositions, possibly spanning from one pure substance to another.

### A1.2.3. *Convention (II)*

Convention (II) (known as the infinitely-dilute solution reference) distinguishes, amongst the components of the solution, the substance (or substances) present in high proportion, known as the solvents, and those which are present in smaller proportions – the solutes. The reference state is different for these two categories of components:

– For a solvent, we choose its pure state (in the same state of aggregation as the solution) as a reference, and therefore its chemical potential obeys relation [A1.9]. Convention (II) is identical to convention (I) in the case of a solvent.

– For a solute, the reference state is an imaginary solution in which all the solutes are infinitely dilute. The reference chemical potential is, therefore, that of the solute in that imaginary solution, and is written as $\mu_s^\infty$ for a solute $s$. The activity coefficient of the solute $s$ in that imaginary solution is equal to 1. Thus, for solute $s$, the chemical potential is written as follows, agreeing that the activity tends toward 1 if the molar fraction tends toward zero:

$$\mu_s = \mu_s^\infty + \mathrm{R}T \ln \gamma_s^{(\mathrm{II})} x_s \qquad \text{[A1.10]}$$

### A1.2.4. *Relation between conventions (I) and (II)*

The activity coefficients of a solute in reference (I) (pure substance) and reference (II) (infinitely-dilute solution) are linked to one another because the chemical potential of the solute is independent of the convention chosen. Therefore, we can write:

$$\mu_s = g_s^0 + \mathrm{R}T \ln \gamma_s^{(\mathrm{I})} x_s = \mu_s^\infty + \mathrm{R}T \ln \gamma_s^{(\mathrm{II})} x_s \qquad \text{[A1.11]}$$

If we let $K_{iH}$ represent *Henry's constant*, defined by the relation:

$$\ln K_{iH} = \frac{\mu_s^\infty - g_s^0}{\mathrm{R}T} \qquad \text{[A1.12]}$$

from this, we deduce:

$$\ln \frac{\gamma_s^{(\mathrm{I})}}{\gamma_s^{(\mathrm{II})}} = \frac{\mu_s^\infty - g_s^0}{\mathrm{R}T} = \ln K_{iH} \qquad \text{[A1.13]}$$

Hence, the constant $K_{iH}$ links the activity coefficients expressed, for a solute, in the two conventions: pure-substance reference and dilute-solution reference. This constant does not depend on the composition of the solution, but instead depends on the values of the intensive variables (pressure, temperature, etc.), by way of the chemical potentials of the reference states, amongst other things.

We have seen that, for the solvent, the two conventions are identical.

NOTE.– In a solution with more than two components, there is no reason not to consider multiple solvents and multiple solutes.

### A1.2.5. *Convention (III)*

Convention (III) again makes the distinction between the solvent and solutes:

– for the solvent, the reference (III) convention is identical to reference (I), the pure-substance reference, and thus the chemical potential of the solvent is always given by relation [A1.9];

– for a solute, we choose to write that the activity coefficient of the solute is equal to 1 for a solution containing 1 mole per liter of each solute. The chemical potential of the reference state is that of the solute in that solution which contains 1 mol/l of each of the solutes. The chemical potential of the solute is therefore written:

$$\mu_s = \mu_s^{0(\text{III})} + \text{R}T \ln \gamma_s^{(\text{III})} C_s \qquad \text{[A1.14]}$$

NOTE.– It is not unfeasible, in certain cases of low solubility, that the reference solution with 1 mol/l of each solute is perfectly imaginary, because it is impossible to realize. This takes nothing away from the convention, which requires only that the solvent and the solutes in the reference solution be in the same state of aggregation as the solution, even if it is a fictitious state.

Convention (III), the molar solution reference, is often used for ionic aqueous solutions, although in the latter case, the chosen reference is often the solution whose molarity tends toward zero, which does not change the chemical potential of the reference state $\mu_s^{0(\text{III})}$.

### A1.2.6. *Relation between conventions (II) and (III)*

Let us examine the relation between the activity coefficients expressed on the basis of the dilute solution reference (II), and those stemming from the molar solution (convention (III)).

Let us write that the chemical potential does not depend on the chosen reference state, and taking account of relation [A1.5], we find:

$$\mu_s = \mu_s^{0(\mathrm{III})} + \mathrm{R}T \ln \gamma_s^{(\mathrm{III})} C_s \cong \mu_s^{\infty} + \mathrm{R}T \ln \gamma_s^{(\mathrm{II})} x_s \qquad [A1.15]$$

Thus, we can identify and write:

$$\mu_s^{0(\mathrm{III})} = \mu_s^{\infty} \qquad [A1.16]$$

We find:

$$\gamma_s^{(\mathrm{III})} C_s = \gamma_s^{(\mathrm{II})} x_s \qquad [A1.17]$$

Thus, the chemical potential in infinite dilution is equal to that at the concentration of 1 mol/l.

The product of the activity coefficient in convention (II) by the molar fraction (i.e. the activity in reference (II)) is equal to the product of the activity coefficient in convention (III) by the concentration expressed in moles per liter (i.e. the activity in reference (III)).

If the solution is very dilute, from relations [A1.17] and [A1.5], we deduce:

$$\gamma_s^{(\mathrm{II})} = \frac{\gamma_s^{(\mathrm{III})}}{v_0^0} \qquad [A1.18]$$

The molar volume of the solvent $v_0^0$ is expressed in liters per mole.

If the concentrations are sufficiently low for us to write that $\gamma_s^{(\mathrm{II})} = 1$, then $\gamma_s^{(\mathrm{III})} = v_0^0$.

### A1.2.7. *Influence of temperature on the reference chemical potentials*

We shall base our discussion on chemical systems.

The expression [A1.7] of the chemical potential of a component in the solution can be written as:

$$\frac{\mu_i}{T} = \frac{\mu_i^0}{T} + \mathrm{R}\ln\gamma_i + \mathrm{R}\ln x_i \qquad \text{[A1.19]}$$

However, in view of Helmholtz's second relation, we have:

$$\frac{\partial\left(\frac{\mu_i}{T}\right)}{\partial T} = -\frac{\overline{H_i}}{T^2} \qquad \text{[A1.20]}$$

Similarly, we are able to apply this relation to the reference state of the solution, as follows:

$$\frac{\partial\left(\frac{\mu_i^0}{T}\right)}{\partial T} = -\frac{\overline{H_i^0}}{T^2} \qquad \text{[A1.21]}$$

$\overline{H_i^0}$ is the partial molar enthalpy of the component in the reference state, at temperature $T$.

This expression can be integrated if we know the variations of the enthalpy with the temperature. Usually, we consider, for finite temperature ranges, that the enthalpy is constant.

### A1.2.8. *Influence of pressure on the reference chemical potentials*

If we choose an intensive variable $Y$, which is the conjugate of the extensive variable $X$, we know that the variation of the generalized chemical potential of a component with that variable $Y$ is of the form:

$$\frac{\partial\mu_i}{\partial Y} = -\overline{X_i} \qquad \text{[A1.22]}$$

$\overline{X_i}$ is the partial molar value corresponding to variable $X$.

Let us now apply this relation to the reference solution. We obtain:

$$\frac{\partial \mu_i^0}{\partial Y} = -\overline{X_i^0} \qquad [A1.23]$$

Such a relation applies, for example, to the influence of the pressure. Remembering that variable $Y$ is –P, we find:

$$\frac{\partial \mu_i^0}{\partial P} = \overline{V_i^0} \qquad [A1.24]$$

$\overline{V_i^0}$ is the partial molar volume of the component in the reference solution. In the case of reference (I), it is simply the molar volume of the pure substance $v_i^0$.

## A1.3. Characterization of the imperfection of a real solution by the excess Gibbs energy

A second technique, which is being used increasingly widely, to define a solution is to define its difference from the perfect solution by the excess Gibbs energy.

### A1.3.1. *Definition of the excess values*

Consider an extensive property $J$ of a solution and call $J^*$ the value that the property would have in the given conditions of temperature and pressure if that solution were perfect. The excess value of $J$ is the value $J^{xs}$ defined by:

$$J^{xs} = J - J^* \qquad [A1.25]$$

$J^{xs}$ is indeed characteristic of the difference of our real solution in comparison to a perfect solution.

As this property is extensive, it has corresponding partial molar values, relating to each component of the solution, defined by:

$$\overline{J^{xs}} = \left( \frac{\partial J^{xs}}{\partial n_i} \right)_{P,T,n_{j \neq i}} \qquad \text{[A1.26]}$$

We can divide this definition for the Gibbs energy.

### A1.3.2. *Excess Gibbs energy*

Consider a solution. The excess Gibbs energy, according to the definition, will be:

$$G^{xs} = G - G^* \qquad \text{[A1.27]}$$

However, the molar Gibbs energy of the solution can be written:

$$G_m = \sum_{i=1}^{N} x_i \mu_i = \sum_{i=1}^{N} \left[ x_i \left( g_i^0 + \mathrm{R}T \ln x_i + \mathrm{R}T \ln \gamma_i^{(\mathrm{I})} \right) \right] \qquad \text{[A1.28]}$$

Similarly, for a perfect solution, this molar Gibbs energy would be:

$$G_m^* = \sum_{i=1}^{N} \left[ x_i \left( g_i^0 + \mathrm{R}T \ln x_i \right) \right] \qquad \text{[A1.29]}$$

From this, we deduce the molar excess Gibbs energy of the solution at hand:

$$G_m^{xs} = \mathrm{R}T \sum_{i=1}^{N} x_i \ln \gamma_i^{(\mathrm{I})} \qquad \text{[A1.30]}$$

If the excess Gibbs energy is known, the solution is completely characterized: if we add the term of the Gibbs energy of the perfect solution, we obtain the Gibbs energy of the solution, which is the characteristic function with our choice of variables $P$, $T$ and composition.

By applying relation [A1.26] to the Gibbs energy, then immediately for the partial molar excess Gibbs energy of the component:

$$\overline{G_i^{xs}} = \mathrm{R}T\left(\ln \gamma_i^{(\mathrm{I})} + \sum_{i=1}^{N} x_j \frac{\partial \ln \gamma_j^{(\mathrm{I})}}{\partial n_i}\right) \qquad [A1.31]$$

However, the Gibbs–Duhem relation enables us to write:

$$\sum_{i=1}^{N} x_j \frac{\partial \ln \gamma_j^{(\mathrm{I})}}{\partial n_i} = 0 \qquad [A1.32]$$

Thus, ultimately, the partial molar excess Gibbs energy of the component is:

$$\overline{G_i^{xs}} = \mathrm{R}T \ln \gamma_i^{(\mathrm{I})} \qquad [A1.33]$$

Relation [A1.33] forms the link between the description of the solutions by the Lewis activity coefficients method, and that by the excess Gibbs energy. The two methods, therefore, are equivalent.

# Appendix 2

## Recap of statistical thermodynamics

The aim of this appendix is to recap, but not demonstrate, certain results in the field of statistical thermodynamics, which are used in this fourth volume in this series. These concepts are presented in detail and demonstrated in the first volume: *Phase Modeling Tools: Applications to Gases* [SOU 15b].

We know that in the microscopic approach to a phase, we consider that phase to be a collection of molecule-objects whose energies are distributed in accordance with a statistical law. The state of a collection of molecule-objects changes constantly, but over time, the collection reaches a certain type of distribution in which the molecule-objects may be found in different states. The term *number of complexions* denotes the number of distributions of molecule-objects between the different states they are liable to occupy. Out of all the possible distributions, there is one which corresponds to the maximum number of complexions. Boltzmann's law allows that the number of complexions corresponding to the most probable type of distribution is practically equal to the total number of complexions, and *vice versa.* The state of the collection is then that which corresponds to the maximum number of complexions.

Most calculations in statistical thermodynamics are based on Stirling's approximation, which enables us to simplify the expression of the factorial logarithm $n$ if the number $n$ is large. It is written:

$$\ln n! \cong n \ln n - n \cong n \ln n \qquad [A2.1]$$

## A2.1. The three branches of statistics

Each element in the collection has an energy $\varepsilon_i$ and the number of elements which has that energy is $n_i$. The total number of elements will be $N$, such that:

$$N = \sum_i n_i \qquad [A2.2]$$

Thus, the total energy is:

$$E = \sum_i n_i \varepsilon_i \qquad [A2.3]$$

The number of complexions, i.e. the number of configurations of the set of elements, is written as Ω.

The mean energy of an element is $\langle \varepsilon \rangle$, and by applying relation [2.2], we find:

$$\langle \varepsilon \rangle = \frac{E}{N} \qquad [A2.4]$$

Thus, for the number of objects in the state $i$, we obtain:

$$n_i = g_i \exp(-\alpha)\exp(-\beta\varepsilon_i) \qquad [A2.5]$$

$g_i$ is the statistical weight or the coefficient of degeneracy or multiplicity of the energy level $\varepsilon_i$: it is the number of different states which have the same energy $\varepsilon_i$.

The coefficient $\beta$ is a universal, whose value is:

$$\beta = \frac{1}{\mathrm{k_B}T} \qquad [A2.6]$$

$\mathrm{k_B}$ is Boltzmann's constant (quotient of the joule gas constant, R, by Avogadro's number $\mathrm{N_a}$).

Depending on the nature of the molecule-object, three branches of statistics can be applied.

### A2.1.1. *Maxwell–Boltzmann statistics*

Maxwell–Boltzmann statistics is applicable to objects for which there is no need to draw on quantum mechanics – i.e. objects which are relatively large and discernible. This type of statistics is also applicable for discernible or localized quantum objects, such as the *molecules* distributed at the nodes of a crystalline lattice.

The coefficient $\alpha$ in relation [A2.5] is given, in this case, by:

$$\exp(-\alpha) = \frac{N}{\sum_i g_i \exp(-\beta\varepsilon_i)} \qquad \text{[A2.7]}$$

The distribution law becomes:

$$n_i = \frac{N g_i \exp(-\beta\varepsilon_i)}{\sum_i g_i \exp(-\beta\varepsilon_i)} \qquad \text{[A2.8]}$$

$\beta$ is still defined by relation [A2.6].

### A2.1.2. *Bose–Einstein quantum statistics*

Bose–Einstein quantum statistics applies for non-localized quantum objects – that is, objects which are indiscernible and have integer spin (such as most molecules and ions, atoms). The distribution of the objects obeys the expression:

$$(n_i)_{BE} = \frac{g_i \exp(-\alpha - \beta\varepsilon_i)}{1 - \exp(-\alpha - \beta\varepsilon_i)} \qquad \text{[A2.9]}$$

The value of the coefficient $\alpha$ is difficult to determine. We shall come back to this later on (relation [A2.11]).

### A2.1.3. *Fermi–Dirac quantum statistics*

Fermi–Dirac quantum statistics applies for non-localized quantum objects – i.e. those which are indiscernible and have fraction spin (some molecules and ions, electrons). The distribution of the objects obeys the expression:

$$(n_i)_{BE} = \frac{g_i \exp(-\alpha - \beta\varepsilon_i)}{1 + \exp(-\alpha - \beta\varepsilon_i)} \quad \text{[A2.10]}$$

The value of the coefficient $\alpha$ is just as difficult to calculate as in the previous case.

For the two branches of statistics pertaining to non-discernible objects, we content ourselves, as regards that coefficient $\alpha$, with a limited expansion of the following form:

$$\exp(-\alpha_{\mathrm{A}}) = a_0 + a_1 \frac{N_{\mathrm{A}}}{z_{\mathrm{A}}} + a_2 \left(\frac{N_{\mathrm{A}}}{z_{\mathrm{A}}}\right)^2 + a_3 \left(\frac{N_{\mathrm{A}}}{z_{\mathrm{A}}}\right)^3 + \ldots \quad \text{[A2.11]}$$

By laborious calculations, it can be shown that the coefficients $a_{\mathrm{i}}$ in that expansion are:

$$a_0 = 0\,;\; a_1 = 1\,;\; a_2 = \pm\frac{1}{2^{3/2}}\,;\; a_3 = \frac{1}{4} - \frac{1}{3^{3/2}}\,;\; \text{etc.} \quad \text{[A2.12]}$$

In the coefficient $a_2$, the + sign is applied for Fermi–Dirac statistics, and the – sign for Bose–Einstein statistics.

### A2.1.4. *Classic limiting case*

The three branches of quantum statistics (Maxwell-Boltzmann, Bose–Einstein and Fermi–Dirac) meld into one, known as the classic limiting case, if the following condition is met:

$$\exp(-\alpha) << 1 \quad \text{[A2.13]}$$

Here, the value of $\alpha$ must be very high.

In these conditions, the three laws are combined in the form:

$$n_i = g_i \exp(-\alpha - \beta\varepsilon_i) \quad \text{[A2.14]}$$

We can see that this limiting case keeps the formula of the Maxwell–Boltzmann distribution. The value of $\alpha$ is therefore determined by the relation:

$$\exp(-\alpha_{\mathrm{A}}) \cong \frac{N_{\mathrm{A}}}{z_{\mathrm{A}}} \qquad [A2.15]$$

We can show that in this case. We also have the condition:

$$N << g_i \qquad [A2.16]$$

This means that the number of particles is much less than the number of possible states.

## A2.2. Partition functions of a molecule object

### A2.2.1. *Definition*

The partition function of molecule object of a collection is the sum $z$ defined by:

$$z = \sum_i g_i \exp\left(-\frac{\varepsilon_i}{\mathrm{k_B}T}\right) = \sum_i g_i \exp(-\varepsilon_i \beta) \qquad [A2.17]$$

The sum is extended to all the energy levels which that object can attain.

### A2.2.2. *Independence of energy values*

The complete partition function for a system includes terms which refer to the different forms of energy: nuclear, electronic, molecule vibration, rotation, translation and the energy of interaction between the different molecules.

For simplicity's sake, we accept the hypothesis that those different forms of energy, for a molecule, are independent.

In these conditions, we can write the total energy of a molecule as the sum of the different contributions of the forms of energy: nuclear $\varepsilon_n$, electronic $\varepsilon_e$, vibrational $\varepsilon_v$, rotational $\varepsilon_r$, translational $\varepsilon_t$ and interactional $\varepsilon_I$, so:

$$\varepsilon = \varepsilon_n + \varepsilon_e + \varepsilon_v + \varepsilon_r + \varepsilon_t + \varepsilon_I \qquad [A2.18]$$

The partition function of the molecule becomes:

$$z = \sum_{i_n} \exp\left(-\frac{\varepsilon_{i_n}}{\mathrm{k_B}T}\right) \sum_{i_e} \exp\left(-\frac{\varepsilon_{i_e}}{\mathrm{k_B}T}\right) \sum_{i_v} \exp\left(-\frac{\varepsilon_{i_v}}{\mathrm{k_B}T}\right) \sum_{i_r} \exp\left(-\frac{\varepsilon_{i_r}}{\mathrm{k_B}T}\right) \sum_{i_t} \exp\left(-\frac{\varepsilon_{i_t}}{\mathrm{k_B}T}\right) \sum_{i_I} \exp\left(-\frac{\varepsilon_{i_I}}{\mathrm{k_B}T}\right) \qquad [A2.19]$$

Thus, we see the emergence of partial partition functions, which pertain to the different forms of energy:

$$z_n = \sum_{i_n} \exp\left(-\frac{\varepsilon_{i_n}}{\mathrm{k_B}T}\right), z_e = \sum_{i_e} \exp\left(-\frac{\varepsilon_{i_e}}{\mathrm{k_B}T}\right), z_v = \sum_{i_v} \exp\left(-\frac{\varepsilon_{i_v}}{\mathrm{k_B}T}\right)$$

$$z_r = \sum_{i_r} \exp\left(-\frac{\varepsilon_{i_r}}{\mathrm{k_B}T}\right), z_t = \sum_{i_t} \exp\left(-\frac{\varepsilon_{i_t}}{\mathrm{k_B}T}\right), z_I = \sum_{i_I} \exp\left(-\frac{\varepsilon_{i_I}}{\mathrm{k_B}T}\right) \qquad [A2.20]$$

The global partition function then takes the form of a product of the partial partition functions:

$$z = z_n z_e z_v z_r z_t z_I \qquad [A2.21]$$

Sometimes, we use the term *internal contribution* to speak of the product:

$$z_{\mathrm{int}} = z_n z_e z_v z_r \qquad [A2.22]$$

This is the product of all the contributions other than those of translation and interaction.

The global molecular partition function then becomes:

$$z = z_{\mathrm{int}} z_t z_I \qquad [A2.23]$$

### A2.2.3. *Partial molecular partition functions, pertaining to the different motions*

By applying the definition [A2.20], we can calculate the contributions of each of the motions of the molecule to the molecular partition function.

#### A2.2.3.1. *Translation*

The molecule has three translational degrees of freedom. We can show that if it is not subjected to any constraint other than the obligation to remain within the volume $V$, the translational contribution is:

$$z_t = V\left(\frac{2\pi m \mathrm{k_B} T}{\mathrm{h}^2}\right)^{3/2} \quad [A2.24]$$

For the perfect gas, with no interaction ($z_I = 1$) between the molecules, the molecular partition function can be written as follows, in view of relation [A2.24]:

$$z_{pf} = V\left(\frac{2\pi m \mathrm{k_B} T}{\mathrm{h}^2}\right)^{3/2} z_{\mathrm{int}} \quad [A2.25]$$

Thus, for the translational partition function of the perfect gas, we have:

$$z_{t(pf)} = V\left(\frac{2\pi m \mathrm{k_B} T}{\mathrm{h}^2}\right)^{3/2} \quad [A2.26]$$

#### A2.2.3.2. *Rotation*

A molecule may have two or three rotational degrees of freedom. We distinguish three categories of molecules.

For homonuclear diatomic molecules (i.e. where the two atoms are identical), the partition function for each degree of freedom is:

$$z_r = \frac{4\pi^2 I \mathrm{k_B} T}{\mathrm{h}^2} \quad [A2.27]$$

For heteronuclear diatomic molecules, the partition function for each degree of freedom is:

$$z_r = \frac{4\pi^2 I k_B T}{h^2} \qquad [A2.28]$$

In the case of complex molecules, the partition function for each degree of freedom is:

$$z_r = \frac{8\pi^2 (2\pi k_B T)^{3/2} \sqrt{I_1 I_2 I_3}}{\sigma h^2} \qquad [A2.29]$$

$\sigma$ is a symmetry coefficient which depends on the complexity of the molecule, and whose value is, at most, a few units.

#### A2.2.3.3. *Vibration of frequency* $\nu$

We can show that the partition function of a degree of freedom with the frequency $\nu$ is given by:

$$z_\nu = \frac{\exp\left[-\frac{h\nu}{2k_B T}\right]}{1-\exp\left[-\frac{h\nu}{k_B T}\right]} \qquad [A2.30]$$

Note that if $\frac{h\nu}{k_B T} >> 1$, we can content ourselves with a simpler formula:

$$z_\nu \approx \frac{k_B T}{h\nu} \qquad [A2.31]$$

### A2.3. Canonical partition function

To use phase-characterization statistics, it is helpful to use the canonical set.

### A2.3.1. *Canonical set*

A *canonical set* is a set composed of replicas of the system under study. Each element is enclosed, so the number of molecules $N$ is identical in all the elements of the set. That number is constant, because there is no exchange of matter between the elements and between the elements and the outside of the set. The volume $V$ is the same for all the elements. The elements are in thermal contact with one another, and are therefore able to exchange energy. Their temperature is identical $T$. Each element has an energy level $E_j$. The total energy of the canonical set will be $E_C$. That energy is constant, because the set is insulated from the outside world.

### A2.3.2. *Canonical partition functions*

In the same way as for molecules, we define the partition function for the canonical set by the sum:

$$Z_C = \sum_j \exp\left(-\beta E_j\right) \qquad [A2.32]$$

This sum is extended to all the elements of the set.

### A2.3.3. *Canonical partition function and molecular partition functions*

The canonical partition function is linked, firstly to the molecular canonical functions, and secondly to the thermodynamic functions that define the phase on the macroscopic level ($U$, $F$, $G$, $S$, etc.). These two types of relation mean that the canonical partition function forms the link between the microscopic definition of the phase and its macroscopic thermodynamic properties.

In order to calculate the canonical partition function on the basis of the molecular functions, we distinguish two cases, depending on whether the molecules are discernible or indiscernible.

#### A2.3.3.1. *Case of sets of discernible molecules*

If the molecules are all identical and discernible, we can show that the following expression can be used:

$$Z_C = z^N \qquad [A2.33]$$

If the phase is made up of several different types of molecules – $N_A$ molecules of A, $N_B$ molecules of B, etc. – then the canonical partition function becomes:

$$Z_C = z_A^{N_A} . z_B^{N_B} .... \qquad [A2.34]$$

### A2.3.3.2. *Case of sets of indiscernible molecules*

If the molecules are all identical and indiscernible, then we can show that the following expression can be employed:

$$Z_C = \frac{z^N}{N\,!} \qquad [A2.35]$$

If the phase is made up of several different types of molecules – $N_A$ molecules of A, $N_B$ molecules of B, etc., the canonical partition function becomes:

$$Z_C = \frac{z_A^{N_A}}{N_A\,!} . \frac{z_B^{N_B}}{N_B\,!} ... = \prod_A \frac{z_A^{N_A}}{N_A\,!} \qquad [A2.36]$$

In the case that the medium under study is a mixture of several phases α, β, γ, etc., the canonical partition function of the set is the product of the canonical partition functions of the different phases, in obeisance to:

$$Z_C = Z_C^{(\alpha)} . Z_C^{(\beta)} . Z_C^{(\gamma)} .. \qquad [A2.37]$$

Each canonical partition function of each of the phases obeys one of relations [A2.33], [A2.34], [A2.35] or [A2.36].

## A2.4. Canonical partition functions and thermodynamic functions

Let us look at the following expressions, used to express the thermodynamic functions on the basis of the canonical partition function.

For the internal energy:

$$U(T)-U(0)=-\sum_{\mathrm{A}}\frac{\partial \ln Z_{C(\mathrm{A})}}{\partial \beta}=\mathrm{k_B}T^2\sum_{\mathrm{A}}\frac{\partial \ln Z_{C(\mathrm{A})}}{\partial T} \qquad [A2.38]$$

For the entropy:

$$S=\mathrm{k_B}\left[\ln Z_C-\frac{1}{T}\frac{\partial \ln Z_C}{\partial \ln \beta}\right] \qquad [A2.39]$$

For the Helmholtz energy function:

$$F(T)-F(0)=-\frac{\sum_{\mathrm{A}}\ln Z_{C(\mathrm{A})}}{\beta}=-\mathrm{k_B}T\sum_{\mathrm{A}}\ln Z_{C(\mathrm{A})} \qquad [A2.40]$$

For the pressure:

Using equation [A2.40], we calculate:

$$P=-\left(\frac{\partial F}{\partial V}\right)_{T,N_A,..}=\mathrm{k_B}T\sum_{\mathrm{A}}\left(\frac{\partial \ln Z_{C(\mathrm{A})}}{\partial V_{\mathrm{A}}}\right)_{T,N_{\mathrm{A}},..} \qquad [A2.41]$$

For the Gibbs energy:

Equations [A2.40] and [A2.41] enable us to write:

$$G-G(0)=F-F(0)-PV=-\frac{\sum_{\mathrm{A}}\ln Z_{C(\mathrm{A})}}{\beta}+\frac{V}{\beta}\sum_{\mathrm{A}}\left(\frac{\partial \ln Z_{C(\mathrm{A})}}{\partial V_{\mathrm{A}}}\right)_{T,N} \qquad [A2.42]$$

NOTE ON THE MOLAR PROPERTIES.– in the expressions of the molecular and canonical partition functions, we have based our discussions on a certain number of molecules – a number of particles symbolized by $N_{\mathrm{A}}$ for the component A.

In order to obtain the molar values of the thermodynamic functions, for $N_{\mathrm{A}}$ it is prudent to choose Avogadro's number ($\mathrm{N_a}$).

To obtain the value of a function for a quantity $n_A$ (in moles) of the component, for $N_A$ we use the product $n_A$ $N_a$.

## A2.5. Equilibrium constants and molecular partition functions

As the thermodynamic constants and, in particular, the Gibbs energies, can be expressed on the basis of the partition functions, the same must be true for the equilibrium constants.

We shall now express the equilibrium constant of a reaction by using the partition functions of the reagents and the products.

Let us consider a chemical reaction between the reagents $A_r$, giving rise to the products $A_p$, which can be written in the form:

$$0 = \sum_p a_p A_p + \sum_r a_r A_r = \sum_i a_i A_i$$

This formulation uses the algebraic stoichiometric numbers $a_i$, which are positive for a product of the reaction and negative for a reagent.

In view of relation [A2.42] and, taking account of the expression of the pressure given by relation [A2.41], we can write the Gibbs energy of a component $i$ in the form:

$$G_i(T) - G_i(0) = -k_B T \ln Z_{iC} + P_i V \qquad \text{[A2.43]}$$

$G_i(0)$ is the Gibbs energy of the pure component $i$ at the temperature of 0 K.

Here, we shall look at two types of equilibrium: homogeneous equilibria in the gaseous phase and homogeneous equilibria in the liquid phase.

### A2.5.1. *Homogeneous equilibria in the gaseous phase*

Gas molecules are considered to be indiscernible molecules, so the canonical partition function is given by relation [A2.36]. When we apply this relation and use Stirling's first approximation [A2.1], relation [A2.43] becomes:

$$G_i(T)-G_i(0)=-N_i \mathrm{k_B} T\ln z_i+\mathrm{k_B}T\left(N_i\ln N_i-N_i\right)+P_iV \qquad \text{[A2.44]}$$

We shall use molar properties. In order to do so, we define a fraction of molar partition $z_{i(m)}$ given by:

$$z_{i(m)}=\frac{z_i}{n_i} \qquad \text{[A2.45]}$$

This definition is tantamount to replacing the volume $V$ in the calculation of the translational partition function with the molar volume $v_{i(m)}$.

In standard conditions, at temperature $T$, the Gibbs energy of $n$ moles of the component $i$, which is a supposedly-perfect gas ($P_iV = n_i\mathrm{R}T$), becomes:

$$g_i^0(T)=g_i^0(0)-n_i\mathrm{R}T\ln\frac{z_{i(m)}}{\mathrm{N_a}} \qquad \text{[A2.46]}$$

For the reaction, we write:

$$\Delta_r g_i^0(T)=\sum_i a_i g_i^0(T)=\sum_i a_i g_i^0(0)-\mathrm{R}T\sum_i a_i\ln\frac{z_{i(m)}}{\mathrm{N_a}} \qquad \text{[A2.47]}$$

However, at the temperature of 0K, for a perfect gas, we have:

$$g_i^0(0)=u_i^0(0)+PV-TS=h_i^0(0)\cong u_i^0(0) \qquad \text{[A2.48]}$$

Consequently, relation [A2.47] becomes:

$$\Delta_r g_i^0(T)=\Delta_r u^0(T)-\mathrm{R}T\sum_i a_i\ln\frac{z_{i(m)}}{\mathrm{N_a}} \qquad \text{[A2.49]}$$

The term $\Delta_r u^0(T)$ is the linear combination, weighted by the stoichiometric numbers, of the residual energies of vibration of each of the substances. Therefore, if each component has $k_i$ vibrational degrees of freedom with the fundamental frequency $\nu^0_{k_i}$, we have:

$$\Delta_r u^0(0) = \frac{\mathrm{N_a h}}{2} \sum_i \left(a_i \sum_k \nu^0_{k_i}\right) \qquad [A2.50]$$

The equilibrium constant is defined by:

$$-RT \ln K_p = \Delta_r g^0(T) \qquad [A2.51]$$

In view of relation [A2.49], we obtain:

$$K_P = \prod_i \left(\frac{z_{i(m)}}{\mathrm{N_a}}\right)^{a_i} \exp - \frac{\Delta_r u^0(0)}{RT} \qquad [A2.52]$$

Remember that this relation is valid for equilibria in the gaseous phase, involving perfect gases.

### A2.5.2. *Homogeneous equilibria in the liquid phase*

If the reaction occurs in the liquid phase, then all the components involved in the reaction are components of a single liquid solution.

For such a solution, the variation in volume due to the reaction is negligible, and therefore we have:

$$\Delta_r G(T) = \Delta_r F(T) + P\Delta_r V \cong \Delta_r F(T) \qquad [A2.53]$$

We can express the Helmholtz energy by relation [A2.40]. The gas molecules are considered as indiscernible molecules, so the canonical partition function is given by relation [A2.36]. Using Stirling's second approximation [A2.1], we see relation [A2.53] become:

$$F_i(T) - F_i(0) = -N_i \mathrm{k_B} T \left(\ln z_i + \ln N_i\right) \qquad [A2.54]$$

In this expression, we suppose we are dealing with a perfect solution, as the equation does not take account of an enthalpy of mixing.

In standard conditions, at temperature $T$, the Helmholtz energy of $n$ moles of component $i$ becomes:

$$f_i^0(T) = f_i^0(0) - n_i \mathrm{R}T \ln \frac{z_{i(m)}}{\mathrm{N_a}} \qquad \text{[A2.55]}$$

For the reaction, we write:

$$\Delta_r f_i^0(T) = \sum_i a_i f_i^0(0) = -\mathrm{R}T \sum_i a_i \ln \frac{z_{i(m)}}{\mathrm{N_a}} \qquad \text{[A2.56]}$$

$f_i^0(0)$ is the standard molar Helmholtz energy of the pure component $i$ at the temperature of 0 K, and $z_{i(m)}$ is its molar partition function.

Using relations [A2.51] and [A2.54], the equilibrium constant is written as:

$$-\mathrm{R}T \ln K_x = \Delta_r f^0(T) \qquad \text{[A2.57]}$$

Using relations [A2.47] and [A2.56], we find:

$$-\mathrm{R}T \ln K_x = \Delta_r \left[u_i^0(0)\right] - \mathrm{R}T \ln \prod_i \left(\frac{z_{i(m)}}{\mathrm{N_a}}\right)^{a_i} \qquad \text{[A2.58]}$$

For the equilibrium constant:

$$K_x = \prod_i \left(\frac{z_{i(m)}}{\mathrm{N_a}}\right)^{a_i} \exp\left(-\frac{\Delta_r u_0}{\mathrm{R}T}\right) \qquad \text{[A2.59]}$$

This relation, which is valid in the case of perfect or highly-dilute solutions, exhibits the same form as relation [A2.52], obtained for a homogeneous reaction between perfect gases.

Certain authors introduce the molecular partition function, defined by:

$$z^*_{i(m)} = \frac{z_{i(m)}}{\mathrm{N_a}} \qquad \text{[A2.60]}$$

Equation [A2.59] then takes on the form:

$$K_x = \prod_i \left(z^*_{i(m)}\right)^{a_i} \exp\left(-\frac{\Delta_r u_0}{RT}\right) \qquad \text{[A2.61]}$$

For solutions exhibiting deviation from ideal behavior, we need to add to the "zero internal energy" the internal energy of mixing, calculated by the relation:

$$\Delta_r U^{mix} = \sum_i a_i \overline{u}_i^{mix} \qquad \text{[A2.62]}$$

and relation [A2.60] therefore takes the form:

$$K_x = \prod_i \left(z^*_{i(m)}\right)^{a_i} \exp\left(-\frac{\Delta_r u_0 + \Delta_r U^{mix}}{RT}\right) \qquad \text{[A2.63]}$$

### A2.5.3. *Homogeneous equilibria in the solid phase*

If the reaction takes place in the solid phase, then all the components in the reaction are components of the same solid solution.

For such a solution, as in the case of liquids, the variation in volume due to the reaction is negligible, and therefore we have:

$$\Delta_r G(T) = \Delta_r F(T) + P\Delta_r V \cong \Delta_r F(T) \qquad \text{[A2.64]}$$

We can express the Helmholtz energy by relation [A2.40]. The molecules in the solid are considered to be discernible molecules, so the canonical partition function is given by relation [A2.34]. Hence, relation [A2.70] becomes:

$$F_i(T) - F_i(0) = -N_i \mathrm{k_B} T \ln z_i \qquad \text{[A2.65]}$$

In that expression, we suppose the solution to be perfect, because we do not take account of an enthalpy of mixing.

In standard conditions, at temperature $T$, the Helmholtz energy of $n$ moles of the component $i$ becomes:

$$f_i^0(T) = f_i^0(0) - n_i RT \ln z_i \qquad \text{[A2.66]}$$

NOTE.– in the case of solids, we do not need to make use of the molar partition functions, because the partition functions of solids contain only vibrational terms, which do not involve the volume of the chamber.

For the reaction, we write:

$$\Delta_r f_i^0(T) = \sum_i a_i f_i^0(0) = -RT \sum_i a_i \ln z_i \qquad [A2.67]$$

$f_i^0(0)$ is the standard molar Helmholtz energy of the pure component $i$ at the temperature of 0 K, and $z$ is its partition function.

Using relations [A2.51] and [A2.651], the equilibrium constant is written:

$$-RT \ln K_x = \Delta_r f^0(T) \qquad [A2.68]$$

Using relations [A2.56] and [A2.66], we find

$$-RT \ln K_x = \Delta_r \left[ u_i^0(0) \right] - RT \ln \prod_i (z_i)^{a_i} \qquad [A2.69]$$

For the equilibrium constant:

$$K_x = \prod_i (z_i)^{a_i} \exp\left( -\frac{\Delta_r u_0}{RT} \right) \qquad [A2.70]$$

Thus, the equilibrium constant in the solid phase can be calculated on the basis of the molecular partition functions of vibration – i.e. the vibration frequencies of the molecules. Those values also enable us to calculate the residual energy of these molecules, and hence, in the case of perfect solutions, to determine the exponential term in relation [A2.70].

# Bibliography


[ABR 95] ABRAMS D.S., PRAUSNITZ J.M., *AICHE Journal*, vol. 21, no. 1, p. 116, 1995.

[ATK 90] ATKINS P.W., *Physical Chemistry*, Oxford University Press, Oxford, 1990.

[BEN 12] BEN-NAIM A., *Entropy and the Second Law*, World Scientific Publishing Company, New Jersey, 2012.

[BER 97] BERNACHE-ASSOLLANT D., COURNIL M., *Equilibres en solution*, Masson, Paris, 1997.

[BES 84] BESSON J., *Précis de thermodynamique et de cinétique électrochimiques*, Ellipses, Paris, 1984.

[DEF 51] DEFAY R., PRIGOGINE I., *Tension superficielle et adsorption*, Edition Desoer, Liège, 1951.

[DEM 73] DEMANGE H., GERMAIN G., NOTIN M., *Comprendre et appliquer la thermodynamique*, Masson, Paris, 1973.

[DEN 68] DENBIGH K., *The Principles of Chemical Equilibrium*, Cambridge University Press, Cambridge, 1968.

[DES 10] DESRE P., HODAJ F., *Thermodynamique des matériaux, Equilibres de phase et métastabilité*, EDP Sciences, Paris, 2010.

[EMS 51] EMSCHWILLER G., *Chimie Physique*, PUF, Paris, 1951.

[EYR 61] EYRING H., REE T., *Proceedings of the National Academy of Sciences of the United States of America*, vol. 47, 1961.

[FOW 49] FOWLER R., GUGGENHEIM E.A., *Statistical Thermodynamics*, Cambridge University Press, London, 1949.

[GER 59] GERMAIN J.E., *Catalyse hétérogène*, Dunod, Paris, 1959.

[GIL 63] GILBERT J., *Chimie-Physique I. Atomistique et liaison chimique*, Masson, Paris, 1963.

[GLA 41] GLASSTONE S.N., LAIDLER K.J., EYRING H.W., *The Theory of Rate Processes*, McGraw-Hill, New York, 1941.

[GOK 96] GOKCEN N.A., REDDY R.G., *Thermodynamics*, Plenum Press, New York, 1996.

[GRE 99] GREINER W., *Mécanique quantique. Une introduction*, Springer, Berlin, 1999.

[GUG 32] GUGGENHEIM E.A., *Proceedings of the Royal Society of London A*, vol. 135, p. 181, 1932.

[HIL 56] HILL T.L., *Statistical Mechanics*, McGraw-Hill, New York, 1956.

[INF 06] INFELTA P., GRAETZEL M., *Thermodynamique, principes et applications*, BrownWalker Press, Boca Raton, FL, 2006.

[KLO 08] KLOTZ I.M., ROSENBERG R.W., *Chemical Thermodynamics: Basic Concepts and Methods*, 7th ed., Wiley & Sons, New York, 2008.

[KRÖ 74] KRÖGER F.A., *The Chemistry of Imperfect Crystals*, North-Holland Publishing Company, Amsterdam, 1974.

[LAL 06] LALAUZE R., *Physico-chimie des interfaces solide-gaz 1*, Hermes Science-Lavoisier, Paris, 2006.

[LEG 05] LE GORREC B., *Thermodynamique et cinétique électrochimique expérimentale*, available at www.minatec.grenoble-inp.fr/Cours-A.pdf, 2005.

[LEN 37] LENNARD-JONES J.E., DEVONSHIRE A.F., *Proceedings of the Royal Society of London A*, vol. 163, p. 53, 1937.

[MAR 95] MARCHAND A., PACAULT A., *La thermodynamique mot à mot*, De Boeck Université, Bruxelles, 1995.

[MCQ 99] MCQUARRIEN D.A., SIMON J.D., *Molecular Thermodynamics*, University Science Books, Sausalito, 1999.

[MEZ 46] MEZGER E., "Loi de variation de la tension superficielle avec la temperature", *J. Phys. Radium*, vol. 7, no. 10, pp. 303–309, 1946.

[MIE 03] MIE G., *Ann. Physik.* vol. 11, p. 657, 1903.

[ONO 47] ONO S., *Mem. Fac. Eng. Kyushu Univ.*, vol. 10, p. 190, 1947.

[PAC 63] PACAULT A., *Eléments de thermodynamique statistique*, Masson, Paris, 1963.

[PRI 46] PRIGOGINE I., DEFAY R., *Thermodynamique chimique*, Dunod, Paris, 1946.

[PRI 66] PRINCE A., *Alloy Phase Equilibria*, Elsevier Publishing Company, Amsterdam, 1966.

[REE 64] REE T.S., REE T., EYRING H., *Proceedings of the National Academy of Sciences of the United States of America*, vol. 51, 1964.

[REI 87] REID R.C., PRAUSNITZ J.M., POLING B.E., *Properties of Gas and Liquids*, McGraw-Hill, New York, 1987.

[REN 71] RENON H., *Calcul sur ordinateur des équilibres liquide, vapeur et liquide-liquide*, Editions Technip, Paris, 1971.

[SOR 81] SORENSEN O.T., *Non-Stoichiometric Oxides*, Academic Press, New York, 1981.

[SOU 68] SOUCHAY P., *Thermodynamique chimique*, Masson, Paris, 1968.

[SOU 73] SOUSTELLE M., GARDET J.J., GUJLHOT B., "Diagrammes de phases et stœchiométrie", in SUCHET J.P. (ed.)., *Diagrammes de phases et stœchiométrie*, Masson, Paris, pp. 33–50, 1973.

[SOU 90] SOUSTELLE M., *Modélisation macroscopique des transformations physico-chimiques*, Masson, Paris, 1990.

[SOU 93] SOUSTELLE M., "Défauts ponctuels dans les solides", in BERNACHE-ASSOLLANT D. (ed.), *Chimie, physique du frittage*, Hermes Science-Lavoisier, Paris, pp. 21–90, 1993.

[SOU 10] SOUSTELLE M., *Handbook of Heterogeneous Kinetics*, ISTE, London and John Wiley & Sons, New York, 2010.

[SOU 11] SOUSTELLE M., *An Introduction to Chemical kinetics*, ISTE, London and John Wiley & Sons, New York, 2011.

[SOU 15a] SOUSTELLE M., *Thermodynamic Modeling of Solid Phases*, ISTE, London and John Wiley & Sons, New York, 2015.

[SOU 15b] SOUSTELLE M., *Phase Modeling Tools: Application to Gases*, ISTE, London, John Wiley & Sons, New York, 2015.

[SOU 15c] SOUSTELLE M., *Modeling of Liquid Phases,* ISTE, London and John Wiley & Sons, New York, 2015.

# Index

## F, G, H, I, J



## K, L, M, O, P



## R, S



## T, V

Other titles from

ISTE

in

Chemical Engineering

## 2015

SOUSTELLE Michel
*Chemical Thermodynamics Set*
*Volume 1 – Phase Modeling Tools*
*Volume 2 – Modeling of Liquid Phases*
*Volume 3 – Thermodynamic Modeling of Solid Phases*

## 2014

DAL PONT Jean-Pierre, AZZARO-PANTEL Catherine
*New Approaches to the Process Industries: The Manufacturing Plant of the Future*

## 2012

DAL PONT Jean-Pierre
*Process Engineering and Industrial Management*

## 2011

SOUSTELLE Michel
*An Introduction to Chemical Kinetics*

## 2010

SOUSTELLE Michel
*Handbook of Heterogenous Kinetics*

Printed and bound by CPI Group (UK) Ltd, Croydon, CR0 4YY
13/09/2026
14961020-0001